D1073047

Discrete Mathematics
Demystified

Demystified Series

Discrete Mathematics
Demystified

Steven G. Krantz

New York   Chicago   San Francisco   Lisbon   London
Madrid   Mexico City   Milan   New Delhi   San Juan
Seoul   Singapore   Sydney   Toronto

The **McGraw·Hill** Companies

Library of Congress Cataloging-in-Publication Data

Krantz, Steven G. (Steven George), date.
   Discrete mathematics demystified / Steven G. Krantz.—1st ed.
      p.   cm.
   ISBN 978-0-07-154948-6 (alk. paper)
   1. Computer science—Mathematics.   I. Title.
QA76.9.M35K74   2009
511′.1—dc22   2008029580

1   2   3   4   5   6   7   8   9   0   DOC/DOC   0   1   4   3   2   1   0   9   8

ISBN 978-0-07-154948-6
MHID 0-07-154948-X

Sponsoring Editor
   Judy Bass

Production Supervisor
   Pamela A. Pelton

Editing Supervisor
   Stephen M. Smith

Project Manager
   Rasika Mathur, International Typesetting
   and Composition

Copy Editor
   Priyanka Sinha, International Typesetting
   and Composition

Proofreader
   Nigel O'Brien, International Typesetting
   and Composition

Indexer
   Brenda Miller

Art Director, Cover
   Jeff Weeks

Composition
   International Typesetting and Composition

Printed and bound by RR Donnelley.

To the memory of J. W. T. Youngs.

ABOUT THE AUTHOR

Steven G. Krantz, Ph.D., is a professor of mathematics at Washington University in St. Louis, Missouri. He just finished a stint as deputy director at the American Institute of Mathematics. Dr. Krantz is an award-winning teacher, and the author of *How to Teach Mathematics, Calculus Demystified,* and *Differential Equations Demystified,* among other books.

CONTENTS

Contents

Contents

PREFACE

In today's world, analytical thinking is a critical part of any solid education. An important segment of this kind of reasoning—one that cuts across many disciplines—is discrete mathematics. Discrete math concerns counting, probability, (sophisticated forms of) addition, and limit processes over discrete sets. Combinatorics, graph theory, the idea of function, recurrence relations, permutations, and set theory are all part of discrete math. Sequences and series are among the most important applications of these ideas.

Discrete mathematics is an essential part of the foundations of (theoretical) computer science, statistics, probability theory, and algebra. The ideas come up repeatedly in different parts of calculus. Many would argue that discrete math is the most important component of all modern mathematical thought.

Most basic math courses (at the freshman and sophomore level) are oriented toward problem-solving. Students can rely heavily on the provided examples as a crutch to learn the basic techniques and pass the exams. Discrete mathematics is, by contrast, rather theoretical. It involves proofs and ideas and abstraction. Freshman and sophomores in college these days have little experience with theory or with abstract thinking. They simply are not intellectually prepared for such material.

Steven G. Krantz is an award-winning teacher, author of the book *How to Teach Mathematics*. He knows how to present mathematical ideas in a concrete fashion that students can absorb and master in a comfortable fashion. He can explain even abstract concepts in a hands-on fashion, making the learning process natural and fluid. Examples can be made tactile and real, thus helping students to finesse abstract technicalities. This book will serve as an ideal supplement to any standard text. It will help students over the traditional "hump" that the first theoretical math course constitutes. It will make the course palatable. Krantz has already authored two successful *Demystified* books.

The good news is that discrete math, particularly sequences and series, *can* be illustrated with concrete examples from the real world. They *can* be made to be realistic and approachable. Thus the rather difficult set of ideas can be made accessible to a broad audience of students. For today's audience—consisting not

only of mathematics students but of engineers, physicists, premedical students, social scientists, and others—this feature is especially important.

A typical audience for this book will be freshman and sophomore students in the mathematical sciences, in engineering, in physics, and in any field where analytical thinking will play a role. Today premedical students, nursing students, business students, and many others take some version of calculus or discrete math or both. They will definitely need help with these theoretical topics.

This text has several key features that make it unique and useful:

1. The book makes abstract ideas concrete. All concepts are presented succinctly and clearly.

2. Real-world examples illustrate ideas and make them accessible.

3. Applications and examples come from real, believable contexts that are familiar and meaningful.

4. Exercises develop both routine and analytical thinking skills.

5. The book relates discrete math ideas to other parts of mathematics and science.

Discrete Mathematics Demystified explains this panorama of ideas in a step-by-step and accessible manner. The author, a renowned teacher and expositor, has a strong sense of the level of the students who will read this book, their backgrounds and their strengths, and can present the material in accessible morsels that the student can study on his or her own. Well-chosen examples and cognate exercises will reinforce the ideas being presented. Frequent review, assessment, and application of the ideas will help students to retain and to internalize all the important concepts of calculus.

Discrete Mathematics Demystified will be a valuable addition to the self-help literature. Written by an accomplished and experienced teacher, this book will also aid the student who is working without a teacher. It will provide encouragement and reinforcement as needed, and diagnostic exercises will help the student to measure his or her progress.

CHAPTER 1

Logic

Strictly speaking, our approach to logic is "intuitive" or "naïve." Whereas in ordinary conversation these emotion-charged words may be used to downgrade the value of that which is being described, our use of these words is more technical. What is meant is that we shall prescribe in this chapter certain rules of logic, which are to be followed in the rest of the book. They will be presented to you in such a way that their validity should be intuitively appealing and self-evident. We cannot *prove* these rules. The rules of logic are the point where our learning begins. A more advanced course in logic will explore other logical methods. The ones that we present here are universally accepted in mathematics and in most of science and analytical thought.

We shall begin with sentential logic and elementary connectives. This material is called the *propositional calculus* (to distinguish it from the predicate calculus, which will be treated later). In other words, we shall be discussing *propositions*—which are built up from atomic statements and connectives. The elementary connectives include "and," "or," "not," "if-then," and "if and only if." Each of these will have a precise meaning and will have exact relationships with the other connectives.

An *elementary statement* (or *atomic statement*) is a sentence with a subject and a verb (and sometimes an object) but no connectives (and, or, not, if then, if, and

only if). For example,

John is good

Mary has bread

Ethel reads books

are all atomic statements. We build up sentences, or propositions, from atomic statements using connectives.

Next we shall consider the quantifiers "for all" and "there exists" and their relationships with the connectives from the last paragraph. The quantifiers will give rise to the so-called *predicate calculus*. Connectives and quantifiers will prove to be the building blocks of all future statements in this book, indeed in all of mathematics.

1.1 Sentential Logic

In everyday conversation, people sometimes argue about whether a statement is true or not. In mathematics there is nothing to argue about. In practice a sensible statement in mathematics is either true or false, and there is no room for opinion about this attribute. How do we determine which statements are true and which are false?

The modern methodology in mathematics works as follows:

- We *define* certain terms.
- We *assume* that these terms, or statements about them, have certain properties or truth attributes (these assumptions are called axioms).
- We *specify* certain rules of logic.

Any statement that can be derived from the axioms, using the rules of logic, is understood to be true. It is not necessarily the case that every true statement can be derived in this fashion. However, in practice this is our method for verifying that a statement is true.

On the other hand, a statement is false if it is inconsistent with the axioms and the rules of logic. That is to say, a statement is false if the assumption that it is true leads to a contradiction. Alternatively, a statement **P** is false if the negation of **P** can be established or proved. While it is possible for a statement to be false without our being able to derive a contradiction in this fashion, in practice we establish falsity by the method of contradiction or by giving a counterexample (which is another aspect of the method of contradiction).

The point of view being described here is special to mathematics. While it is indeed true that mathematics is used to model the world around us—in physics, engineering, and in other sciences—the subject of mathematics itself is a man-made system. Its internal coherence is guaranteed by the axiomatic method that we have just described.

It is worth mentioning that "truth" in everyday life is treated differently. When you tell someone "I love you" and you are asked for proof, a mathematical verification will not do the job. You will offer empirical evidence of your caring, your fealty, your monogamy, and so forth. But you cannot give a mathematical proof. In a court of law, when an attorney "proves" a case, he/she does so by offering evidence and arguing from that evidence. The attorney *cannot* offer a mathematical argument.

The way that we reason in mathematics is special, but it is ideally suited to the task that we must perform. It is a means of rigorously manipulating ideas to arrive at new truths. It is a methodology that has stood the test of time for thousands of years, and that guarantees that our ideas will travel well and apply to a great variety of situations and applications.

It is reasonable to ask whether mathematical truth is a construct of the human mind or an immutable part of nature. For instance, is the assertion that "the area of a circle is π times the radius squared" actually a fact of nature just like Newton's inverse square law of gravitation? Our point of view is that mathematical truth is relative. The formula for the area of a circle is a logical consequence of the axioms of mathematics, nothing more. The fact that the formula seems to describe what is going on in nature is convenient, and is part of what makes mathematics useful. But that aspect is something over which we as mathematicians have no control. Our concern is with the internal coherence of our logical system.

It can be asserted that a "proof" (a concept to be discussed later in the book) is a psychological device for convincing the reader that an assertion is true. However, our view in this book is more rigid: a proof of an assertion is a sequence of applications of the rules of logic to derive the assertion from the axioms. There is no room for opinion here. The axioms are plain. The rules are rigid. A proof is like a sequence of moves in a game of chess. If the rules are followed then the proof is correct. Otherwise not.

1.2 "And" and "Or"

Let **A** be the statement "Arnold is old." and **B** be the statement "Arnold is fat." The new statement

"A   and   B"

means that both **A** is true *and* **B** is true. Thus

Arnold is old and Arnold is fat

means both that Arnold is old *and* Arnold is fat. If we meet Arnold and he turns out to be young and fat, then the statement is false. If he is old and thin then the statement is false. Finally, if Arnold is *both* young and thin then the statement is false. The statement is *true* precisely when both properties—oldness and fatness—hold. We may summarize these assertions with a *truth table*. We let

A = Arnold is old

and

B = Arnold is fat

The expression

A ∧ B

will denote the phrase "**A and B**." We call this statement the *conjunction* of **A** and **B**. The letters "T" and "F" denote "True" and "False" respectively. Then we have

A	B	A ∧ B
T	T	T
T	F	F
F	T	F
F	F	F

Notice that we have listed all possible truth values of **A** and **B** and the corresponding values of the *conjunction* **A** ∧ **B**. The conjunction is true only when *both* **A** and **B** are true. Otherwise it is false. This property is a special feature of conjunction, or "and."

In a restaurant the menu often contains phrases such as

soup or salad

This means that we may select soup *or* select salad, but we may not select both. This use of the word "or" is called the *exclusive* "or"; it is not the meaning of "or" that we use in mathematics and logic. In mathematics we instead say that "**A** or **B**" is true provided that **A** is true or **B** is true or *both* are true. This is the *inclusive* "or." If we let **A** ∨ **B** denote "**A** or **B**" then the truth table is

A	B	A ∨ B
T	T	T
T	F	T
F	T	T
F	F	F

We call the statement **A** ∨ **B** the *disjunction* of **A** and **B**. Note that this disjunction is true in three out of four cases: the only time the disjunction is false is if both components are false.

The reason that we use the inclusive form of "or" in mathematics is that this form of "or" has a nice relationship with "and," as we shall see below. The other form of "or" does not.

We see from the truth table that the only way that "**A or B**" can be false is if *both* **A** is false and **B** is false. For instance, the statement

Hilary is beautiful or Hilary is poor

means that Hilary is either beautiful or poor or both. In particular, she will not be both ugly and rich. Another way of saying this is that if she is ugly she will compensate by being poor; if she is rich she will compensate by being beautiful. *But she could be both beautiful and poor.*

EXAMPLE 1.1
The statement

$$x > 2 \quad \text{and} \quad x < 5$$

is true for the number $x = 3$ because this value of x is both greater than 2 *and* less than 5. It is false for $x = 6$ because this x value is greater than 2 but not less than 5. It is false for $x = 1$ because this x is less than 5 but not greater than 2.   □

EXAMPLE 1.2
The statement

x is odd and x is a perfect cube

is true for $x = 27$ because both assertions hold. It is false for $x = 7$ because this x, while odd, is not a cube. It is false for $x = 8$ because this x, while a cube, is not odd. It is false for $x = 10$ because this x is neither odd nor is it a cube.   □

EXAMPLE 1.3
The statement

$$x < 3 \quad \text{or} \quad x > 6$$

is true for $x = 2$ since this x is < 3 (even though it is not > 6). It holds (that is, it is true) for $x = 9$ because this x is > 6 (even though it is not < 3). The statement fails (that is, it is false) for $x = 4$ since this x is neither < 3 nor > 6. □

EXAMPLE 1.4
The statement

$$x > 1 \quad \text{or} \quad x < 4$$

is true for every real x. As an exercise, you should provide a detailed reason for this answer. (**Hint:** Consider separately the cases $x < 1, x = 1, 1 < x < 4, x = 4$, and $x > 4$.) □

EXAMPLE 1.5
The statement $(\mathbf{A} \vee \mathbf{B}) \wedge \mathbf{B}$ has the following truth table:

A	B	A ∨ B	(A ∨ B) ∧ B
T	T	T	T
T	F	T	F
F	T	T	T
F	F	F	F

□

You Try It: Construct a truth table for the statement

The number x is positive and is a perfect square

Notice in Example 1.5 that the statement $(\mathbf{A} \vee \mathbf{B}) \wedge \mathbf{B}$ has the same truth values as the simpler statement $\mathbf{B}$. In what follows, we shall call such pairs of statements (having the same truth values) *logically equivalent*.

The words "and" and "or" are called *connectives*; their role in sentential logic is to enable us to build up (or to connect together) pairs of statements. The idea is to use very simple statements, like "**Jennifer is swift.**" as building blocks; then we compose more complex statements from these building blocks by using connectives.

In the next two sections we will become acquainted with the other two basic connectives "not" and "if-then." We shall also say a little bit about the compound connective "if and only if."

1.3 "Not"

The statement "not **A**," written $\sim \mathbf{A}$, is true whenever **A** is false. For example, the statement

Charles is not happily married

is true provided the statement "Charles is happily married" is false. The truth table for $\sim \mathbf{A}$ is as follows:

A	$\sim$ A
T	F
F	T

Greater understanding is obtained by combining the connectives:

EXAMPLE 1.6
We examine the truth table for $\sim (\mathbf{A} \wedge \mathbf{B})$:

A	B	A $\wedge$ B	$\sim (\mathbf{A} \wedge \mathbf{B})$
T	T	T	F
T	F	F	T
F	T	F	T
F	F	F	T

$\square$

EXAMPLE 1.7
Now we look at the truth table for $(\sim \mathbf{A}) \vee (\sim \mathbf{B})$:

A	B	$\sim$ A	$\sim$ B	$(\sim \mathbf{A}) \vee (\sim \mathbf{B})$
T	T	F	F	F
T	F	F	T	T
F	T	T	F	T
F	F	T	T	T

$\square$

Notice that the statements $\sim (A \wedge B)$ and $(\sim A) \vee (\sim B)$ have the *same truth table*. As previously noted, such pairs of statements are called logically equivalent.

The logical equivalence of $\sim (A \wedge B)$ with $(\sim A) \vee (\sim B)$ makes good intuitive sense: the statement $A \wedge B$ fails [that is, $\sim (A \wedge B)$ is true] precisely when either A is false *or* B is false. That is, $(\sim A) \vee (\sim B)$. Since in mathematics we cannot rely on our intuition to establish facts, it is important to have the truth table technique for establishing logical equivalence. The exercise set will give you further practice with this notion.

One of the main reasons that we use the inclusive definition of "or" rather than the exclusive one is so that the connectives "and" and "or" have the nice relationship just discussed. It is also the case that $\sim (A \vee B)$ and $(\sim A) \wedge (\sim B)$ are logically equivalent. These logical equivalences are sometimes referred to as *de Morgan's laws*.

1.4 "If-Then"

A statement of the form "If A then B" asserts that whenever A is true then B is also true. This assertion (or "promise") is tested when A is true, because it is then claimed that something else (namely B) is true as well. *However*, when A is false then the statement "If A then B" *claims nothing*. Using the symbols $A \Rightarrow B$ to denote "If A then B", we obtain the following truth table:

A	B	$A \Rightarrow B$
T	T	T
T	F	F
F	T	T
F	F	T

Notice that we use here an important principle of aristotelian logic: every sensible statement is either true or false. There is no "in between" status. When A is false we can hardly assert that $A \Rightarrow B$ is false. For $A \Rightarrow B$ asserts that "whenever A is true then B is true", and A is not true!

Put in other words, when A is false then the statement $A \Rightarrow B$ is not tested. It therefore cannot be false. So it must be true. We refer to A as the *hypothesis* of the implication and to B as the *conclusion* of the implication. When the if-then statement is true, then the hypothsis implies the conclusion.

EXAMPLE 1.8
The statement "If $2 = 4$ then Calvin Coolidge was our greatest president" is true. This is the case no matter what you think of Calvin Coolidge. The point is that

the hypothesis $(2 = 4)$ is false; thus it doesn't matter what the truth value of the conclusion is. According to the truth table for implication, the sentence is true.

The statement "If fish have hair then chickens have lips" is true. Again, the hypothesis is false so the sentence is true.

The statement "If $9 > 5$ then dogs don't fly" is true. In this case the hypothesis is certainly true and so is the conclusion. Therefore the sentence is true.

(Notice that the "if" part of the sentence and the "then" part of the sentence need not be related in any intuitive sense. The truth or falsity of an "if-then" statement is simply a fact about the logical values of its hypothesis and of its conclusion.) □

EXAMPLE 1.9

The statement $\mathbf{A} \Rightarrow \mathbf{B}$ is logically equivalent with $(\sim \mathbf{A}) \vee \mathbf{B}$. For the truth table for the latter is

A	**B**	$\sim \mathbf{A}$	$(\sim \mathbf{A}) \vee \mathbf{B}$
T	T	F	T
T	F	F	F
F	T	T	T
F	F	T	T

which is the same as the truth table for $\mathbf{A} \Rightarrow \mathbf{B}$. □

You should think for a bit to see that $(\sim \mathbf{A}) \vee \mathbf{B}$ *says the same thing* as $\mathbf{A} \Rightarrow \mathbf{B}$. To wit, assume that the statement $(\sim \mathbf{A}) \vee \mathbf{B}$ is true. Now suppose that $\mathbf{A}$ is true. It follows that $\sim \mathbf{A}$ is false. Then, according to the disjunction, $\mathbf{B}$ must be true. But that says that $\mathbf{A} \Rightarrow \mathbf{B}$. For the converse, assume that $\mathbf{A} \Rightarrow \mathbf{B}$ is true. This means that if $\mathbf{A}$ holds then $\mathbf{B}$ must follow. But that just says $(\sim \mathbf{A}) \vee \mathbf{B}$. So the two statements are equivalent, that is, they say the same thing.

Once you believe that assertion, then the truth table for $(\sim \mathbf{A}) \vee \mathbf{B}$ gives us another way to understand the truth table for $\mathbf{A} \Rightarrow \mathbf{B}$.[1]

There are in fact infinitely many pairs of logically equivalent statements. But just a few of these equivalences are really important in practice—most others are built up from these few basic ones. Some of the other basic pairs of logically equivalent statements are explored in the exercises.

EXAMPLE 1.10

The statement

If x is negative then $-5 \cdot x$ is positive

---

[1] Once again, this logical equivalence illustrates the usefulness of the inclusive version of "or."

is true. For if $x < 0$ then $-5 \cdot x$ is indeed > 0; if $x \geq 0$ then the statement is unchallenged. □

EXAMPLE 1.11
The statement

$$\textbf{If } (x > 0 \textbf{ and } x^2 < 0) \textit{ then } x \geq 10$$

is true since the hypothesis "$x > 0$ **and** $x^2 < 0$" is never true. □

EXAMPLE 1.12
The statement

$$\textbf{If } x > 0 \textbf{ then } (x^2 < 0 \textbf{ or } 2x < 0)$$

is false since the conclusion "$x^2 < 0$ **or** $2x < 0$" is false whenever the hypothesis $x > 0$ is true. □

EXAMPLE 1.13
Let us construct a truth table for the statement $[\textbf{A} \vee (\sim \textbf{B})] \Rightarrow [(\sim \textbf{A}) \wedge \textbf{B}]$.

A	B	$\sim$ A	$\sim$ B	$[\textbf{A} \vee (\sim \textbf{B})]$	$[(\sim \textbf{A}) \wedge \textbf{B}]$
T	T	F	F	T	F
T	F	F	T	T	F
F	T	T	F	F	T
F	F	T	T	T	F

$[\textbf{A} \vee (\sim \textbf{B})] \Rightarrow [(\sim \textbf{A}) \wedge \textbf{B}]$
F
F
T
F

□

Notice that the statement $[\textbf{A} \vee (\sim \textbf{B})] \Rightarrow [(\sim \textbf{A}) \wedge \textbf{B}]$ has the same truth table as $\sim (\textbf{B} \Rightarrow \textbf{A})$. Can you comment on the logical equivalence of these two statements?

Perhaps the most commonly used logical syllogism is the following. Suppose that we know the truth of **A** and of $\textbf{A} \Rightarrow \textbf{B}$. We wish to conclude **B**. Examine the truth table for $\textbf{A} \Rightarrow \textbf{B}$. The only line in which both **A** is true and $\textbf{A} \Rightarrow \textbf{B}$ is true is the line in which **B** is true. That justifies our reasoning. In logic texts, the syllogism we are discussing is known as *modus ponendo ponens* or, more briefly, *modus ponens*.

EXAMPLE 1.14
Consider the two statements

> **It is cloudy**

and

> **If it is cloudy then it is raining**

We think of the first of these as **A** and the second as **A** $\Rightarrow$ **B**. From these two taken together we may conclude **B**, or

> **It is raining**   □

EXAMPLE 1.15
The statement

> **Every yellow dog has fleas**

together with the statement

> **Fido is a blue dog**

allows no logical conclusion. The first statement has the form **A** $\Rightarrow$ **B** but the second statement is *not* **A**. So *modus ponendo ponens* does not apply.   □

EXAMPLE 1.16
Consider the two statements

> **All Martians eat breakfast**

and

> **My friend Jim eats breakfast**

It is quite common, in casual conversation, for people to abuse logic and to conclude that Jim must be a Martian. Of course this is an incorrect application of *modus ponendo ponens*. In fact no conclusion is possible.   □

1.5 Contrapositive, Converse, and "Iff"

The statement

$$\text{If A then B}$$

is the same as

$$A \Rightarrow B$$

or

$$\text{A suffices for B}$$

or as saying

$$\text{A only if B}$$

All these forms are encountered in practice, and you should think about them long enough to realize that they say the same thing.

On the other hand,

$$\text{If B then A}$$

is the same as saying

$$B \Rightarrow A$$

or

$$\text{A is necessary for B}$$

or as saying

$$\text{A if B}$$

We call the statement $B \Rightarrow A$ the *converse* of $A \Rightarrow B$. The converse of an implication is logically distinct from the implication itself. Generally speaking, the converse will *not* be logically equivalent to the original implication. The next two examples illustrate the point.

EXAMPLE 1.17
The converse of the statement

$$\text{If } x \text{ is a healthy horse then } x \text{ has four legs}$$

is the statement

If x has four legs then x is a healthy horse

Notice that these statements have very different meanings: the first statement is true while the second (its converse) is false. For instance, a chair has four legs but it is not a healthy horse. Likewise for a pig. □

EXAMPLE 1.18
The statement

If $x > 5$ then $x > 3$

is true. Any number that is greater than 5 is certainly greater than 3. But the converse

If $x > 3$ then $x > 5$

is certainly false. Take $x = 4$. Then the hypothesis is true but the conclusion is false. □

The statement

A if and only if B

is a brief way of saying

If A then B   and   If B then A

We abbreviate **A if and only if B** as $\mathbf{A} \Leftrightarrow \mathbf{B}$ or as **A iff B**. Now we look at a truth table for $\mathbf{A} \Leftrightarrow \mathbf{B}$.

A	B	$A \Rightarrow B$	$B \Rightarrow A$	$A \Leftrightarrow B$
T	T	T	T	T
T	F	F	T	F
F	T	T	F	F
F	F	T	T	T

Notice that we can say that $\mathbf{A} \Leftrightarrow \mathbf{B}$ is true only when both $\mathbf{A} \Rightarrow \mathbf{B}$ and $\mathbf{B} \Rightarrow \mathbf{A}$ are true. An examination of the truth table reveals that $\mathbf{A} \Leftrightarrow \mathbf{B}$ is true precisely when **A** and **B** are either both true or both false. Thus $\mathbf{A} \Leftrightarrow \mathbf{B}$ means precisely that **A** and **B** are logically equivalent. One is true when and *only when* the other is true.

EXAMPLE 1.19
The statement

$$x > 0 \Leftrightarrow 2x > 0$$

is true. For if $x > 0$ then $2x > 0$; and if $2x > 0$ then $x > 0$. □

EXAMPLE 1.20
The statement

$$x > 0 \Leftrightarrow x^2 > 0$$

is false. For $x > 0 \Rightarrow x^2 > 0$ is certainly true while $x^2 > 0 \Rightarrow x > 0$ is false $[(-3)^2 > 0$ but $-3 \not> 0]$. □

EXAMPLE 1.21
The statement

$$[\sim (A \vee B)] \Leftrightarrow [(\sim A) \wedge (\sim B)] \tag{1.1}$$

is true because the truth table for $\sim(A \vee B)$ and that for $(\sim A) \wedge (\sim B)$ are the same. Thus they are logically equivalent: one statement is true precisely when the other is. Another way to see the truth of Eq. (1.1) is to examine the truth table:

A	B	$\sim (A \vee B)$	$(\sim A) \wedge (\sim B)$	$[\sim (A \vee B)] \Leftrightarrow [(\sim A) \wedge (\sim B)]$
T	T	F	F	T
T	F	F	F	T
F	T	F	F	T
F	F	T	T	T

□

Given an implication

$$A \Rightarrow B$$

the *contrapositive* statement is defined to be the implication

$$\sim B \Rightarrow \sim A$$

The contrapositive (unlike the converse) *is* logically equivalent to the original implication, as we see by examining their truth tables:

A	B	A $\Rightarrow$ B
T	T	T
T	F	F
F	T	T
F	F	T

and

A	B	$\sim$ A	$\sim$ B	($\sim$ B) $\Rightarrow$ ($\sim$ A)
T	T	F	F	T
T	F	F	T	F
F	T	T	F	T
F	F	T	T	T

EXAMPLE 1.22
The statement

> **If it is raining, then it is cloudy**

has, as its contrapositive, the statement

> **If there are no clouds, then it is not raining**

A moment's thought convinces us that these two statements say the same thing: if there are no clouds, then it could not be raining; for the presence of rain implies the presence of clouds.   □

The main point to keep in mind is that, given an implication **A** $\Rightarrow$ **B**, its *converse* **B** $\Rightarrow$ **A** and its *contrapositive* ($\sim$ **B**) $\Rightarrow$ ($\sim$ **A**) are entirely different statements. The converse is distinct from, and *logically independent from*, the original statement. The contrapositive is distinct from, but *logically equivalent to*, the original statement.

Some classical treatments augment the concept of *modus ponens* with the idea of *modus tollendo tollens* or *modus tollens*. It is in fact logically equivalent to *modus ponens*. *Modus tollens* says

> **If $\sim$ B   and   A $\Rightarrow$ B then $\sim$ A**

Modus tollens actualizes the fact that $\sim \mathbf{B} \Rightarrow \sim \mathbf{A}$ is logically equivalent to $\mathbf{A} \Rightarrow \mathbf{B}$. The first of these implications is of course the *contrapositive* of the second.

1.6 Quantifiers

The mathematical statements that we will encounter in practice will use the *connectives* "and," "or," "not," "if-then," and "iff." They will also use *quantifiers*. The two basic quantifiers are "for all" and "there exists".

EXAMPLE 1.23
Consider the statement

All automobiles have wheels

This statement makes an assertion about *all* automobiles. It is true, because every automobile does have wheels.

Compare this statement with the next one:

There exists a woman who is blonde

This statement is of a different nature. It does not claim that all women have blonde hair—merely that there exists *at least one* woman who does. Since that is true, the statement is true. □

EXAMPLE 1.24
Consider the statement

All positive real numbers are integers

This sentence asserts that something is true for all positive real numbers. It is indeed true for *some* positive numbers, such as 1 and 2 and 193. However, it is false for at least one positive number (such as $1/10$ or π), so the entire statement is false.

Here is a more extreme example:

The square of any real number is positive

This assertion is *almost* true—the only exception is the real number 0: $0^2 = 0$ is not positive. But it only takes one exception to falsify a "for all" statement. So the assertion is false.

This last example illustrates the principle that the negation of a "for all" statement is a "there exists" statement. □

EXAMPLE 1.25
Look at the statement

There exists a real number which is greater than 5

In fact there are lots of numbers which are greater than 5; some examples are 7, 42, 2π, and 97/3. Other numbers, such as 1, 2, and $\pi/6$, are not greater than 5. Since there is *at least one* number satisfying the assertion, the assertion is true. □

EXAMPLE 1.26
Consider the statement

There is a man who is at least 10 feet tall

This statement is false. To *verify* that it is false, we must demonstrate that *there does not exist a man who is at least 10 feet tall*. In other words, we must show that all men are shorter than 10 feet.

The negation of a "there exists" statement is a "for all" statement.

A somewhat different example is the sentence

There exists a real number which satisfies the equation
$$x^3 - 2x^2 + 3x - 6 = 0$$

There is in fact only one real number which satisfies the equation, and that is $x = 2$. Yet that information is sufficient to show that the statement true. □

We often use the symbol $\forall$ to denote "for all" and the symbol $\exists$ to denote "there exists." The assertion

$$\forall x, x + 1 < x$$

claims that for every x, the number $x + 1$ is less than x. If we take our universe to be the standard real number system, then this statement is false. The assertion

$$\exists x, x^2 = x$$

claims that there is a number whose square equals itself. If we take our universe to be the real numbers, then the assertion is satisfied by $x = 0$ and by $x = 1$. Therefore the assertion is true.

In all the examples of quantifiers that we have discussed so far, we were careful to specify our *universe* (or at least the universe was clear from context). That is, "There is a woman such that ..." or "All positive real numbers are ..." or "All automobiles have ....." The quantified statement makes no sense unless we specify the universe of objects from which we are making our specification. In the discussion that follows, we will always interpret quantified statements in terms of a universe. Sometimes the universe will be explicitly specified, while other times it will be understood from context.

Quite often we will encounter $\forall$ and $\exists$ used together. The following examples are typical:

EXAMPLE 1.27
The statement

$$\forall x\ \exists y,\ y > x$$

claims that, for any real number x, there is a number y which is greater than it. In the realm of the real numbers this is true. In fact $y = x + 1$ will always do the trick.

The statement

$$\exists x\ \forall y,\ y > x$$

has quite a different meaning from the first one. It claims that there is an x which is less than *every* y. This is absurd. For instance, x is *not* less than $y = x - 1$. $\square$

EXAMPLE 1.28
The statement

$$\forall x\ \forall y,\ x^2 + y^2 \geq 0$$

is true in the realm of the real numbers: it claims that the sum of two squares is always greater than or equal to zero. (This statement happens to be *false* in the realm of the complex numbers. When we interpret a logical statement, it will always be important to understand the context, or universe, in which we are working.)

The statement

$$\exists x \exists y,\ x + 2y = 7$$

is true in the realm of the real numbers: it claims that there exist x and y such that $x + 2y = 7$. Certainly the numbers $x = 3$, $y = 2$ will do the job (although there are many other choices that work as well). $\square$

It is important to note that $\forall$ and $\exists$ do *not* commute. That is to say, $\forall\exists$ and $\exists\forall$ do *not* mean the same thing. Examine Example 1.27 with this thought in mind to make sure that you understand the point.

We conclude by noting that ∀ and ∃ are closely related. The statements

$$\forall x, \mathbf{B}(x) \quad \text{and} \quad \sim \exists x, \sim \mathbf{B}(x)$$

are logically equivalent. The first asserts that the statement $\mathbf{B}(x)$ is true for all values of x. The second asserts that there exists no value of x for which $\mathbf{B}(x)$ fails, which is the same thing.

Likewise, the statements

$$\exists x, \mathbf{B}(x) \quad \text{and} \quad \sim \forall x, \sim \mathbf{B}(x)$$

are logically equivalent. The first asserts that there is some x for which $\mathbf{B}(x)$ is true. The second claims that it is not the case that $\mathbf{B}(x)$ fails for every x, which is the same thing.

A "for all" statement is something like the conjunction of a very large number of simpler statements. For example, the statement

$$\text{For every nonzero integer } n, \ n^2 > 0$$

is actually an efficient way of saying that $1^2 > 0$ and $(-1)^2 > 0$ and $2^2 > 0$, and so on. It is not feasible to apply truth tables to "for all" statements, and we usually do not do so.

A "there exists" statement is something like the disjunction of a very large number of statements (the word "disjunction" in the present context means an "or" statement). For example, the statement

$$\text{There exists an integer } n \text{ such that } \mathbf{P}(n) = 2n^2 - 5n + 2 = 0$$

is actually an efficient way of saying that "$\mathbf{P}(1) = 0$ or $\mathbf{P}(-1) = 0$ or $\mathbf{P}(2) = 0$, and so on." It is not feasible to apply truth tables to "there exist" statements, and we usually do not do so.

It is common to say that *first-order logic* consists of the connectives $\wedge, \vee, \sim, \Rightarrow$, $\Longleftrightarrow$, the equality symbol $=$, and the quantifiers ∀ and ∃, together with an infinite string of variables $x, y, z, \ldots, x', y', z', \ldots$ and, finally, parentheses (, ,) to keep things readable. The word "first" here is used to distinguish the discussion from second-order and higher-order logics. In first-order logic the quantifiers ∀ and ∃ always range over elements of the domain M of discourse. Second-order logic, by contrast, allows us to quantify over subsets of M and functions F mapping $M \times M$ into M. Third-order logic treats sets of function and more abstract constructs. The distinction among these different orders is often moot.

Exercises

1. Construct truth tables for each of the following sentences:

   a. $(S \wedge T) \vee \sim (S \vee T)$

   b. $(S \vee T) \Rightarrow (S \wedge T)$

2. Let

   S = All fish have eyelids.

   T = There is no justice in the world.

   U = I believe everything that I read.

   V = The moon's a balloon.

   Express each of the following sentences using the letters **S, T, U, V** and the connectives $\vee, \wedge, \sim, \Rightarrow, \Leftrightarrow$. *Do not use quantifiers.*

   a. If fish have eyelids then there is at least some justice in the world.

   b. If I believe everything that I read then either the moon's a balloon or at least some fish have no eyelids.

3. Let

   S = All politicians are honest.

   T = Some men are fools.

   U = I don't have two brain cells to rub together.

   W = The pie is in the sky.

   Translate each of the following into English sentences:

   a. $(S \wedge \sim T) \Rightarrow \sim U$

   b. $W \vee (T \wedge \sim U)$

4. State the converse and the contrapositive of each of the following sentences. Be sure to label each.

   a. In order for it to rain it is necessary that there be clouds.

   b. In order for it to rain it is sufficient that there be clouds.

5. Assume that the universe is the ordinary system $\mathbb{R}$ of real numbers. Which of the following sentences is true? Which is false? Give reasons for your answers.

   a. If π is rational then the area of a circle is $E = mc^2$.

   b. If $2 + 2 = 4$ then $3/5$ is a rational number.

6. For each of the following statements, formulate a logically equivalent one using only **S, T,** ~, and ∨. (Of course you may use as many parentheses as you need.) *Use a truth table or other means to explain why the statements are logically equivalent.*

   a. **S** ⇒~ **T**
   b. ~ **S**∧ ~ **T**

7. For each of the following statements, formulate an English sentence that is its negation:

   a. The set *S* contains at least two integers.

   b. Mares eat oats and does eat oats.

8. Which of these pairs of statements is logically equivalent? Why?

   (a) **A**∨ ~ **B**   ~ **A** ⇒ **B**
   (b) **A**∧ ~ **B**   ~ **A** ⇒~ **B**

CHAPTER 2

Methods of Mathematical Proof

2.1 What Is a Proof?

When a chemist asserts that a substance that is subjected to heat will tend to expand, he/she verifies the assertion through experiment. It is a consequence of the *definition* of heat that heat will excite the atomic particles in the substance; it is plausible that this in turn will necessitate expansion of the substance. However, our knowledge of nature is not such that we may turn these theoretical ingredients into a categorical proof. Additional complications arise from the fact that the word "expand" requires detailed definition. Apply heat to water that is at temperature 40 degree Fahrenheit or above, and it expands—with enough heat it becomes a gas that surely fills more volume than the original water. But apply heat to a diamond and there is no apparent "expansion"—at least not to the naked eye.

Mathematics is a less ambitious subject. In particular, it is closed. It does not reach outside itself for verification of its assertions. When we make an assertion in mathematics, we must verify it using the rules that we have laid down. That is, we verify it by applying our rules of logic to our axioms and our definitions; in other words, we construct a *proof*.

In modern mathematics we have discovered that there are perfectly sensible mathematical statements that in fact *cannot* be verified in this fashion, nor can they be proven false. This is a manifestation of Gödel's incompleteness theorem: that any sufficiently complex logical system will contain such unverifiable, indeed untestable, statements. Fortunately, in practice, such statements are the exception rather than the rule. In this book, and in almost all of university-level mathematics, we concentrate on learning about statements whose truth or falsity *is* accessible by way of proof.

This chapter considers the notion of mathematical proof. We shall concentrate on the three principal types of proof: direct proof, proof by contradiction, and proof by induction. In practice, a mathematical proof may contain elements of several or all of these techniques. You will see all the basic elements here. You should be sure to master each of these proof techniques, both so that you can recognize them in your reading and so that they become tools that you can use in your own work.

2.2 Direct Proof

In this section we shall assume that you are familiar with the positive integers, or *natural numbers* (a detailed treatment of the natural numbers appears in Sec. 5.2). This number system $\{1, 2, 3, \ldots\}$ is denoted by the symbol $\mathbb{N}$. For now we will take the elementary arithmetic properties of $\mathbb{N}$ for granted. We shall formulate various statements about natural numbers and we shall prove them. Our methodology will emulate the discussions in earlier sections. We begin with a definition.

Definition 2.1 A natural number n is said to be *even* if, when it is divided by 2, there is an integer quotient and no remainder.

Definition 2.2 A natural number n is said to be *odd* if, when it is divided by 2, there is an integer quotient and remainder 1.

You may have never before considered, at this level of precision, what is the meaning of the terms "odd" or "even." But your intuition should confirm these definitions. A good definition should be precise, but it should also appeal to your heuristic idea about the concept that is being defined.

Notice that, according to these definitions, any natural number is either even or odd. For if n is any natural number, and if we divide it by 2, then the remainder will be either 0 or 1—there is no other possibility (according to the Euclidean algorithm). In the first instance, n is even; in the second, n is odd.

In what follows we will find it convenient to think of an even natural number as one having the form $2m$ for some natural number m. We will think of an odd natural number as one having the form $2k + 1$ for some nonnegative integer k. Check for yourself that, in the first instance, division by 2 will result in a quotient of m and a remainder of 0; in the second instance it will result in a quotient of k and a remainder of 1.

Now let us formulate a statement about the natural numbers and prove it. Following tradition, we refer to formal mathematical statements either as *theorems* or *propositions* or sometimes as *lemmas*. A theorem is supposed to be an important statement that is the culmination of some development of significant ideas. A proposition is a statement of lesser intrinsic importance. Usually a lemma is of no intrinsic interest, but is needed as a step along the way to verifying a theorem or proposition.

Proposition 2.1 *The square of an even natural number is even.*

Proof: Let us begin by using what we learned in Chap. 1. We may reformulate our statement as "If n is even then $n \cdot n$ is even." This statement makes a promise. Refer to the definition of "even" to see what that promise is:

If n can be written as twice a natural number then $n \cdot n$ can be written as twice a natural number.

The hypothesis of the assertion is that $n = 2 \cdot m$ for some natural number m. But then

$$n^2 = n \cdot n = (2m) \cdot (2m) = 4m^2 = 2(2m^2)$$

Our calculation shows that n^2 is twice the natural number $2m^2$. So n^2 is also even.

We have shown that the hypothesis that n is twice a natural number entails the conclusion that n^2 is twice a natural number. In other words, if n is even then n^2 is even. That is the end of our proof.   □

Remark 2.1 What is the role of truth tables at this point? Why did we not use a truth table to verify our proposition? One *could* think of the statement that we are proving as the conjunction of infinitely many specific statements about concrete instances of the variable n; and then we could verify each one of those statements. But such a procedure is inelegant and, more importantly, impractical.

For our purpose, the truth table *tells us what we must do to construct a proof.* The truth table for $\mathbf{A} \Rightarrow \mathbf{B}$ shows that if $\mathbf{A}$ is false then there is nothing to check whereas if **A** is true then we must show that **B** is true. That is just what we did in the proof of Proposition 2.1.

Most of our theorems are "for all" statements or "there exists" statements. In practice, it is not usually possible to verify them directly by use of a truth table.

Proposition 2.2 *The square of an odd natural number is odd.*

Proof: We follow the paradigm laid down in the proof of the previous proposition. Assume that n is odd. Then $n = 2k + 1$ for some nonnegative integer k. But then

$$n^2 = n \cdot n = (2k + 1) \cdot (2k + 1) = 4k^2 + 4k + 1 = 2(2k^2 + 2k) + 1$$

We see that n^2 is $2k' + 1$, where $k' = 2k^2 + 2k$. In other words, according to our definition, n^2 is odd. □

Both of the proofs that we have presented are examples of "direct proof." A direct proof proceeds according to the statement being proved; for instance, if we are proving a statement about a square then we calculate that square. If we are proving a statement about a sum then we calculate that sum. Here are some additional examples:

EXAMPLE 2.1
Prove that, if n is a positive integer, then the quantity $n^2 + 3n + 2$ is even.

Solution: Denote the quantity $n^2 + 3n + 2$ by K. Observe that

$$K = n^2 + 3n + 2 = (n + 1)(n + 2)$$

Thus K is the product of two successive integers: $n + 1$ and $n + 2$. One of those two integers must be even. So it is a multiple of 2. Therefore K itself is a multiple of 2. Hence K must be even. □

You Try It: Prove that the cube of an odd number must be odd.

Proposition 2.3 *The sum of two odd natural numbers is even.*

Proof: Suppose that p and q are both odd natural numbers. According to the definition, we may write $p = 2r + 1$ and $q = 2s + 1$ for some nonnegative integers r and s. Then

$$p + q = (2r + 1) + (2s + 1) = 2r + 2s + 2 = 2(r + s + 1)$$

We have realized $p + q$ as twice the natural number $r + s + 1$. Therefore $p + q$ is even. □

Remark 2.2 If we did mathematics solely according to what sounds good, or what appeals intuitively, then we might reason as follows: "if the sum of two odd natural numbers is even then it must be that the sum of two even natural numbers is odd." This is incorrect. For instance 4 and 6 are each even but their sum $4 + 6 = 10$ is *not* odd.

Intuition definitely plays an important role in the development of mathematics, but all assertions in mathematics must, in the end, be proved by rigorous methods.

EXAMPLE 2.2
Prove that the sum of an even integer and an odd integer is odd.

Solution: An even integer is divisible by 2, so may be written in the form $e = 2m$, where m is an integer. An odd integer has remainder 1 when divided by 2, so may be written in the form $o = 2k + 1$, where k is an integer. The sum of these is

$$e + o = 2m + (2k + 1) = 2(m + k) + 1$$

Thus we see that the sum of an even and an odd integer will have remainder 1 when it is divided by 2. As a result, the sum is odd. □

Proposition 2.4 *The sum of two even natural numbers is even.*

Proof: Let $p = 2r$ and $q = 2s$ both be even natural numbers. Then

$$p + q = 2r + 2s = 2(r + s)$$

We have realized $p + q$ as twice a natural number. Therefore we conclude that $p + q$ is even. □

Proposition 2.5 *Let n be a natural number. Then either $n > 6$ or $n < 9$.*

Proof: If you draw a picture of a number line then you will have no trouble convincing yourself of the truth of the assertion. What we want to learn here is to organize our thoughts so that we may write down a rigorous proof.

Our discussion of the connective "or" in Sec. 1.2 will now come to our aid. Fix a natural number n. If $n > 6$ then the "or" statement is true and there is nothing to prove. If $n \not> 6$, then the truth table teaches us that we must check that $n < 9$. But the statement $n \not> 6$ means that $n \leq 6$ so we have

$$n \leq 6 < 9$$

That is what we wished to prove. □

EXAMPLE 2.3
Prove that every even integer may be written as the sum of two odd integers.

Solution: Let the even integer be $K = 2m$, for m an integer. If m is odd then we write

$$K = 2m = m + m$$

and we have written K as the sum of two odd integers. If, instead, m is even, then we write

$$K = 2m = (m - 1) + (m + 1)$$

Since m is even then both $m - 1$ and $m + 1$ are odd. So again we have written K as the sum of two odd integers. $\square$

EXAMPLE 2.4
Prove the Pythagorean theorem.

Solution: The Pythagorean theorem states that $c^2 = a^2 + b^2$, where a and b are the legs of a right triangle and c is its hypotenuse. See Fig. 2.1.

Consider now the arrangement of four triangles and a square shown in Fig. 2.2. Each of the four triangles is a copy of the original triangle in Fig. 2.1. We see that each side of the all-encompassing square is equal to c. So the area of that square is c^2. Now each of the component triangles has base a and height b. So each such

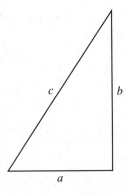

Figure 2.1 The pythagorean theorem.

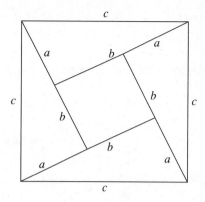

Figure 2.2   Proof of the pythagorean theorem.

triangle has area $ab/2$. And the little square in the middle has side $b - a$. So it has area $(b - a)^2 = b^2 - 2ab + a^2$. We write the total area as the sum of its component areas:

$$c^2 = 4 \cdot \left[\frac{ab}{2} \right] + [b^2 - 2ab + a^2] = a^2 + b^2$$

That is the desired equality. $\square$

In this section and the next two we are concerned with form rather than substance. We are not interested in proving anything profound, but rather in showing you what a proof looks like. Later in the book we shall consider some deeper mathematical ideas and correspondingly more profound proofs.

2.3 Proof by Contradiction

Aristotelian logic dictates that every sensible statement has a truth value: TRUE or FALSE. If we can demonstrate that a statement **A** could not possibly be false, then it must be true. On the other hand, if we can demonstrate that **A** could not be true, then it must be false. Here is a dramatic example of this principle. In order to present it, we shall assume for the moment that you are familiar with the system $\mathbb{Q}$ of rational numbers. These are numbers that may be written as the quotient of two integers (without dividing by zero, of course). We shall discuss the rational numbers in greater detail in Sec. 5.4.

Theorem 2.1 (Pythagoras) *There is no rational number x with the property that $x^2 = 2$.*

Proof: In symbols (refer to Chap. 1), our assertion may be written

$$\sim [\exists x, (x \in \mathbb{Q} \wedge x^2 = 2)]$$

Let us assume the statement to be false. Then what we are assuming is that

$$\exists x, (x \in \mathbb{Q} \wedge x^2 = 2) \tag{2.1}$$

Since x is rational we may write $x = p/q$, where p and q are integers.

We may as well suppose that both p and q are positive and nonzero. After reducing the fraction, we may suppose that it is in lowest terms—so p and q have no common factors.

Now our hypothesis asserts that

$$x^2 = 2$$

or

$$\left(\frac{p}{q}\right)^2 = 2$$

We may write this out as

$$p^2 = 2q^2 \tag{2.2}$$

Observe that this equation asserts that p^2 is an even number. But then p must be an even number (p cannot be odd, for that would imply that p^2 is odd by Proposition 2.2). So $p = 2r$ for some natural number r.

Substituting this assertion into Eq. (2.2) now yields that

$$(2r)^2 = 2q^2$$

Simplifying, we may rewrite our equation as

$$2r^2 = q^2$$

This new equation asserts that q^2 is even. But then q itself must be even.

We have proven that both p and q are even. But that means that they have a common factor of 2. This contradicts our starting assumption that p and q have no common factor.   □

Let us pause to ascertain what we have established: the assumption that a rational square root x of 2 exists, and that it has been written in lowest terms as $x = p/q$, leads to the conclusion that p and q have a common factor and hence are *not* in lowest terms. What does this entail for our logical system?

We cannot allow a statement of the form $\mathbf{C} = \mathbf{A} \wedge \sim \mathbf{A}$ (in the present context the statement $\mathbf{A}$ is "$x = p/q$ in lowest terms"). For such a statement $\mathbf{C}$ must be false.

But if x exists then the statement $\mathbf{C}$ is true. No statement (such as $\mathbf{A}$) can have two truth values. In other words, the statement $\mathbf{C}$ must be false. The only possible conclusion is that x does not exist. That is what we wished to establish.

Remark 2.3 In practice, we do not include the last three paragraphs in a proof by contradiction. We provide them now because this is our first exposure to such a proof, and we want to make the reasoning absolutely clear. The point is that the assertions $\mathbf{A}$ and $\sim \mathbf{A}$ cannot both be true. An assumption that leads to this eventuality cannot be valid. That is the essence of proof by contradiction.

Historically, Theorem 2.1 was extremely important. Prior to Pythagoras (~ 300 B.C.E.), the ancient Greeks (following Eudoxus) believed that all numbers (at least all numbers that arise in real life) are rational. However, by the Pythagorean theorem, the length of the diagonal of a unit square is a number whose square is 2. And our theorem asserts that such a number cannot be rational. We now know that there are many nonrational, or irrational numbers.

Here is a second example of a proof by contradiction:

Theorem 2.2 (Dirichlet) *Suppose that $n + 1$ pieces of mail are delivered to n mailboxes. Then some mailbox contains at least two pieces of mail.*

Proof: Suppose that the assertion is false. Then each mailbox contains either zero or one piece of mail. But then the total amount of mail in all the mailboxes cannot exceed

$$\underbrace{1 + 1 + \cdots + 1}_{n \text{ times}}$$

In other words, there are at most n pieces of mail. That conclusion contradicts the fact that there are $n + 1$ pieces of mail. We conclude that some mailbox contains at least two pieces of mail.   □

Figure 2.3 Points at random in the unit interval.

The last theorem, due to Gustav Lejeune Dirichlet (1805–1859), was classically known as the *Dirichletscher Schubfachschluss*. This German name translates to "Dirichlet's drawer shutting principle." Today, at least in this country, it is more commonly known as "the pigeonhole principle." Since pigeonholes are no longer a common artifact of everyday life, we have illustrated the idea using mailboxes.

EXAMPLE 2.5
Draw the unit interval I in the real line. Now pick 11 points at random from that interval (imagine throwing darts at the interval, or dropping ink drops on the interval). See Fig. 2.3. Then some pair of the points has distance not greater than 0.1 inch.
   To see this, write

$$I = [0, 0.1] \cup [0.1, 0.2] \cup \cdots [0.8, 0.9] \cup [0.9, 1]$$

Here we have used standard interval notation. Think of each of these subintervals as a mailbox. We are delivering 11 letters (that is, the randomly selected points) to these 10 mailboxes. By the pigeonhole principle, some mailbox must receive two letters.
   We conclude that some subinterval of I, having length 0.1, contains two of the randomly selected points. Thus their distance does not exceed 0.1 inch. □

2.4 Proof by Induction

The logical validity of the method of proof by induction is intimately bound up with the construction of the natural numbers, with ordinal arithmetic, and with the so-called well-ordering principle. We shall not treat those logical niceties here, but shall instead concentrate on the technique. As with any good idea in mathematics, we shall be able to make it intuitively clear that the method is a valid and useful one. So no confusion should result.
   Consider a statement $P(n)$ about the natural numbers. For example, the statement might be "The quantity $n^2 + 5n + 6$ is always even." If we wish to prove this statement, we might proceed as follows:

1. Prove the statement $P(1)$.
2. Prove that $P(k) \Rightarrow P(k+1)$ for every $k \in \{1, 2, \ldots\}$.

Let us apply the syllogism *modus ponendo ponens* from the end of Sec. 1.5 to determine what we will have accomplished. We know $P(1)$ and, from Step (**2**) with $k = 1$, that $P(1) \Rightarrow P(2)$. We may therefore conclude $P(2)$. Now Step (**2**) with $k = 2$ says that $P(2) \Rightarrow P(3)$. We may then conclude $P(3)$. Continuing in this fashion, we may establish $P(n)$ for every natural number n.

Notice that this reasoning applies to any statement $P(n)$ for which we can establish Steps (**1**) and (**2**) above. Thus Steps (**1**) and (**2**) taken together constitute a method of proof. It is a method of establishing a statement $P(n)$ for every natural number n. The method is known as *proof by induction*.

EXAMPLE 2.6

Let us use the method of induction to prove that, for every natural number n, the number $n^2 + 5n + 6$ is even.

Solution: Our statement $P(n)$ is

$$\text{The number } n^2 + 5n + 6 \text{ is even}$$

[*Note:* Explicitly identifying $P(n)$ is more than a formality. *Always* record carefully what $P(n)$ is before proceeding.]

We now proceed in two steps:

$P(1)$ is true.   When $n = 1$ then

$$n^2 + 5n + 6 = 1^2 + 5 \cdot 1 + 6 = 12$$

and this is certainly even. We have verified $P(1)$.

$P(n) \Rightarrow P(n+1)$ is true.   We are proving an implication in this step. We *assume* $P(n)$ and *use it* to establish $P(n+1)$. Thus we are assuming that

$$n^2 + 5n + 6 = 2m$$

for some natural number m. Then, to check $P(n+1)$, we calculate

$$(n+1)^2 + 5(n+1) + 6 = [n^2 + 2n + 1] + [5n + 5] + 6$$
$$= [n^2 + 5n + 6] + [2n + 6]$$
$$= 2m + [2n + 6]$$

Notice that in the last step we have *used our hypothesis* that $n^2 + 5n + 6$ is even, that is that $n^2 + 5n + 6 = 2m$. Now the last line may be rewritten as

$$2(m + n + 3)$$

Thus we see that $(n + 1)^2 + 5(n + 1) + 6$ is twice the natural number $m + n + 3$. In other words, $(n + 1)^2 + 5(n + 1) + 6$ is even. But that is the assertion $P(n + 1)$.

In summary, assuming the assertion $P(n)$ we have established the assertion $P(n + 1)$. That completes Step **(2)** of the method of induction. We conclude that $P(n)$ is true for every n. □

Here is another example to illustrate the method of induction. This formula is often attributed to Carl Friedrich Gauss (and alleged to have been discovered by him when he was 11 years old).

Proposition 2.6 *If n is any natural number then*

$$1 + 2 + \cdots + n = \frac{n(n + 1)}{2}$$

Proof: The statement $P(n)$ is

$$1 + 2 + \cdots + n = \frac{n(n + 1)}{2}$$

Now let us follow the method of induction closely.

P(1) is true. The statement $P(1)$ is

$$1 = \frac{1(1 + 1)}{2}$$

This is plainly true.

$P(n) \Rightarrow P(n + 1)$ is true. We are proving an implication in this step. We *assume* $P(n)$ and *use it* to establish $P(n + 1)$. Thus we are assuming that

$$1 + 2 + \cdots + n = \frac{n(n + 1)}{2} \tag{2.3}$$

Let us add the quantity $(n + 1)$ to both sides of Eq. (2.3). We obtain

$$1 + 2 + \cdots + n + (n + 1) = \frac{n(n + 1)}{2} + (n + 1)$$

The left side of this last equation is exactly the left side of $P(n+1)$ that we are trying to establish. That is the motivation for our last step.

Now the right-hand side may be rewritten as

$$\frac{n(n+1) + 2(n+1)}{2}$$

This simplifies to

$$\frac{(n+1)(n+2)}{2}$$

In conclusion, we have established that

$$1 + 2 + \cdots + n + (n+1) = \frac{(n+1)(n+2)}{2}$$

This is the statement $P(n+1)$.

Assuming the validity of $P(n)$, we have proved the validity of $P(n+1)$. That completes the second step of the method of induction, and establishes $P(n)$ for all n.   □

Some problems are formulated in such a way that it is convenient to begin the induction with some value of n other than $n = 1$. The next example illustrates this notion:

EXAMPLE 2.7
Let us prove that, for $n \geq 4$, we have the inequality

$$3^n > 2n^2 + 3n$$

Solution: The statement $P(n)$ is

$$3^n > 2n^2 + 3n$$

$P(4)$ **is true.** Observe that the inequality is false for $n = 1, 2, 3$. However, for $n = 4$ it is certainly the case that

$$3^4 = 81 > 44 = 2 \cdot 4^2 + 3 \cdot 4$$

$P(n) \Rightarrow P(n + 1)$ **is true.** Now assume that $P(n)$ has been established and let us use it to prove $P(n + 1)$. We are hypothesizing that

$$3^n > 2n^2 + 3n$$

Multiplying both sides by 3 gives

$$3 \cdot 3^n > 3(2n^2 + 3n)$$

or

$$3^{n+1} > 6n^2 + 9n$$

But now we have

$$3^{n+1} > 6n^2 + 9n$$
$$= 2(n^2 + 2n + n) + (4n^2 + 3n)$$
$$> 2(n^2 + 2n + 1) + (3n + 3)$$
$$= 2(n + 1)^2 + 3(n + 1)$$

This inequality is just $P(n + 1)$, as we wished to establish. That completes Step (2) of the induction, and therefore completes the proof. □

We conclude this section by mentioning an alternative form of the induction paradigm which is sometimes called *complete mathematical induction* or *strong mathematical induction*.

2.4.1 COMPLETE MATHEMATICAL INDUCTION

Let P be a function on the natural numbers.

1. $P(1)$;
2. $[P(j)$ for all $j \leq n] \Rightarrow P(n + 1)$ for every natural number n;

then $P(n)$ is true for every n.

It turns out that the complete induction principle is logically equivalent to the ordinary induction principle enunciated at the outset of this section. But in some instances strong induction is the more useful tool. An alternative terminology for complete induction is "the set formulation of induction."

Complete induction is sometimes more convenient, or more natural, to use than ordinary induction; it finds particular use in abstract algebra. Complete induction also is a simple instance of transfinite induction, which you will encounter in a more advanced course.

EXAMPLE 2.8
Every integer greater than 1 is either prime or the product of primes. (Here a prime number is an integer whose only factors are 1 and itself.)

Solution: We will use strong induction, just to illustrate the idea. For convenience we begin the induction process at the index 2 rather than at 1.

Let $P(n)$ be the assertion "Either n is prime or n is the product of primes." Then $P(2)$ is plainly true since 2 is the first prime. Now assume that $P(j)$ is true for $2 \leq j \leq n$ and consider $P(n+1)$. If $n+1$ is prime then we are done. If $n+1$ is not prime then $n+1$ factors as $n+1 = k \cdot \ell$, where k, ℓ are integers less than $n+1$, but at least 2. By the strong inductive hypothesis, each of k and ℓ factors as a product of primes (or is itself a prime). Thus $n+1$ factors as a product of primes.

The complete induction is done, and the proof is complete.   □

2.5 Other Methods of Proof

We give here a number of examples that illustrate proof techniques other than direct proof, proof by contradiction, and induction.

2.5.1 COUNTING ARGUMENTS

EXAMPLE 2.9
Show that if there are 23 people in a room then the odds are better than even that two of them have the same birthday.

Solution: The best strategy is to calculate the odds that *no two* people have the same birthday, and then to take complements.

Let us label the people $p_1, p_2, \ldots, p_{23}$. Then, assuming that none of the p_j have the same birthday, we see that p_1 can have his birthday on any of the 365 days in the year, p_2 can then have his birthday on any of the remaining 364 days, p_3 can have his birthday on any of the remaining 363 days, and so forth. So the number of different ways that these 23 people can all have different birthdays is

$$365 \cdot 364 \cdot 363 \cdots 345 \cdot 344 \cdot 343$$

On the other hand, the number of ways that birthdays could be distributed (with no restrictions) among 23 people is

$$\underbrace{365 \cdot 365 \cdot 365 \cdots 365}_{23 \text{ times}} = 365^{23}$$

Thus the probability that these 23 people all have different birthdays is

$$p = \frac{365 \cdot 364 \cdot 363 \cdots 343}{365^{23}}$$

A quick calculation with a calculator shows that $p \sim 0.4927 < 0.5$. That is the desired result. ☐

EXAMPLE 2.10
Jill is dealt a poker hand of 5 cards from a standard deck of 52. What is the probability that she holds four of a kind?

Solution: If the hand holds 4 aces, then the fifth card is any one of the other 48 cards. If the hand holds 4 kings, then the fifth card is any one of the other 48 cards. And so forth. So there are a total of

$$13 \times 48 = 624$$

possible hands with four of a kind. The total number of possible 5-card hands is

$$\binom{52}{5} = 2598960$$

Therefore the probability of holding 4 of a kind is

$$p = \frac{624}{2598960} = 0.00024$$

☐

2.5.2 OTHER ARGUMENTS

EXAMPLE 2.11
Let us show that there exist irrational numbers a and b such that a^b is rational.

Solution: Let $\alpha = \sqrt{2}$ and $\beta = \sqrt{2}$. If α^β is rational then we are done, using $a = \alpha$ and $b = \beta$. If α^β is irrational, then observe that

$$\alpha^{\beta^{\sqrt{2}}} = \alpha^{[\beta \cdot \sqrt{2}]} = \alpha^2 = [\sqrt{2}]^2 = 2$$

Thus, with $a = \alpha^\beta$ and $b = \sqrt{2}$ we have found two irrational numbers a, b such that $a^b = 2$ is rational.   □

EXAMPLE 2.12
Show that if there are six people in a room then either three of them know each other or three of them do not know each other. (Here three people know each other if each of the three pairs has met. Three people do not know each other if each of the three pairs has *not* met.)

Solution: The tedious way to do this problem is to write out all possible "acquaintance assignments" for 6 people.

We now describe a more efficient, and more satisfying, strategy. Call one of the people Bob. There are five others. Either Bob knows three of them, or he does not know three of them.

Say that Bob knows three of the others. If any two of those three are acquainted, then those two and Bob form a mutually acquainted threesome. If no two of those three know each other, then those three are a mutually unacquainted threesome.

Now suppose that Bob does not know three of the others. If any two of those three are unacquainted, then those two and Bob form an unacquainted threesome. If all pairs among the three are instead acquainted, then those three form a mutually acquainted threesome.

We have covered all possibilities, and in every instance come up either with a mutually acquainted threesome or a mutually unacquainted threesome. That ends the proof.   □

It may be worth knowing that five people is insufficient to guarantee either a mutually acquainted threesome or a mutually unacquainted threesome. We leave it to the reader to provide a suitable counterexample. It is quite difficult to determine the minimal number of people to solve the problem when "threesome" is replaced by "foursome." When "foursome" is replaced by five people, the problem is considered to be grossly intractable. This problem is a simple example from the mathematical subject known as *Ramsey theory*.

Exercises

1. Prove that the product of two odd natural numbers must be odd.

2. Prove that if n is an even natural number and if m is *any* natural number then $n \cdot m$ must be even.

3. Prove that the sum of the squares of the first n natural numbers is equal to

$$\frac{2n^3 + 3n^2 + n}{6}$$

4. Prove that if m is a power of 3 and n is a power of 3 then $m + n$ is never a power of 3.

5. Prove that if n is a natural number and if n has a rational square root then in fact the square root of n is an integer.

6. Prove that if the natural number n is a perfect square then $n + 1$ will never be a perfect square.

7. A popular recreational puzzle hypothesizes that you have nine pearls that are identical in appearance. Eight of these pearls have the same weight, but the ninth is either heavier or lighter—you do not know which. You have a balance scale, and are allowed three weighings to find the odd pearl. How do you proceed?

   Now here is a bogus proof by induction that you can solve the problem in the first paragraph in three weighings not just for nine pearls but for any number of pearls. For convenience let us begin the induction with the case $n = 9$ pearls. By the result of the first paragraph, we can handle that case. Now, inductively, suppose that we have an algorithm for handling n pearls. We use this hypothesis to treat $(n + 1)$ pearls. From the $(n + 1)$ pearls, remove one and put it in your pocket. There remain n pearls. Apply the n-pearl algorithm to these remaining pearls. If you find the odd pearl then you are done. If you do not find the odd pearl, then it is the one in your pocket. That completes the case $(n + 1)$ and the proof.

   What is the flaw in this reasoning? [*Note*: If you are fiendishly clever, then you can actually handle 12 pearls in the original problem—with just three weighings. However, this requires the consideration of 27 cases.]

8. Prove that if k is a natural number that is greater than 2 then $2^k > 1 + 2k$.

9. Prove the pigeonhole principle by induction.

10. You write 27 letters to 27 different people. Then you address the 27 envelopes. You close your eyes and stuff one letter into each envelope. What is the probability that just one letter is in the wrong envelope?

CHAPTER 3

Set Theory

3.1 Rudiments

Even the most elementary considerations in logic may lead to conundrums. Suppose that we wish to define the notion of "line." We might say that it is the shortest path between two points. This is not completely satisfactory because we have not yet defined "path" or "point." And when we say "the shortest path" do we mean that there is just one unique shortest path? And why does it exist? Every new definition is, perforce, formulated in terms of other ideas. And those ideas in terms of other ones. Where does the regression cease?

The accepted method for dealing with this problem is to begin with certain terms (as few as possible) that are agreed to be "undefinable." These terms should be so simple that there can be little argument as to their meaning. But it is agreed in advance that these undefinable terms simply cannot be defined in terms of ideas that have been previously defined. Our undefined terms are our starting place.

In modern mathematics it is customary to use "set" and "element of" as undefinables. A *set* is declared to be a collection of objects. (Please do not ask what an "object" is or what a "collection" is; when we say that the term "set" is an undefinable then we mean just that.) If S is a set then we say that x is an element of S, and we write $x \in S$ or $S \ni x$, precisely when x is one of the objects that compose the set S. For example, we write $5 \in \mathbb{N}$ to indicate that the number 5 is an element of the set of natural numbers. We write $-7 \notin \mathbb{N}$ to specify that -7 is *not* an element of the set of natural numbers.

Definition 3.1 We say that two sets S and T are equal precisely when they have the same elements. We write $S = T$. Alternatively, we say that $S = T$ provided that both $x \in S \Rightarrow x \in T$ and $x \in T \Rightarrow x \in S$.

As an example of equality of sets, if $S = \{x \in \mathbb{N} : x^2 > 3\}$ and $T = \{x \in \mathbb{N} : x \geq 2\}$ then $S = T$.

Incidentally, the method of specifying a set with the notation $\{x : P(x)\}$, where P denotes a property, is the most common method in mathematics of defining a set. This is sometimes called "setbuilder notation."

We shall endeavor, in what follows, to formulate all of our set-theoretic notions in a rigorous and logical fashion from the undefinables "set" and "element of." If at any point we were to arrive at an untenable position, or a logical contradiction, or a fallacy,[1] then we know that the fault lies with either our method of reasoning or with our undefinables or with our axioms. One of the advantages of the way that we do mathematics is that if there is ever trouble then we know where the trouble must lie.

3.2 Elements of Set Theory

Beginning in this section, we will be doing mathematics in the way that it is usually done. That is, we shall define terms and we shall state and prove properties that they satisfy. In earlier chapters we were careful, but we were less mathematical. Sometimes we even had to say "This is the way we do it; don't worry." Many of the topics in Chaps. 1 and 2 are really only best understood from the advanced perspectives of mathematical logic. Now, and for the rest of this book, it is time to show how mathematics is done in practice.

We use theorems, propositions, and lemmas to formulate our ideas. The device of proofs is used to validate those ideas. Another formal ingredient of mathematical exposition is the "definition." A definition usually introduces a new piece of terminology or a new idea and *explains what it means in terms of ideas and terminology that have already been presented.* As you read this chapter, pause frequently to check that we are following this paradigm.

---

[1]The reader should not worry. This is well-trodden ground, and there are no pitfalls lurking ahead.

Definition 3.2 Let S and T be sets. We say that S is a *subset* of T, and we write $S \subset T$ or $T \supset S$, if

$$x \in S \Rightarrow x \in T$$

We do not prove our definitions. There is *nothing to prove*. A definition introduces you to a new idea, or piece of terminology, or piece of notation.

EXAMPLE 3.1
Let $S = \{x \in \mathbb{N} : x > 3\}$ and $T = \{x \in \mathbb{N} : x^2 > 4\}$. Determine whether $S \subset T$ or $T \subset S$.

Solution: The key to success and clarity in handling subset questions is to *use the definition*. To see whether $S \subset T$ we must check whether $x \in S$ implies $x \in T$. Now if $x \in S$ then $x > 3$ hence $x^2 > 9$ so certainly $x^2 > 4$. Our syllogism is proved, and we conclude that $S \subset T$.

The reverse inclusion is false. For example, the number 3 is an element of T but is certainly not an element of S. We write $T \not\subset S$.   □

EXAMPLE 3.2
Let $\mathbb{Z}$ denote the set of integers. Let $S = \{-2, 3\}$. Let $T = \{x \in \mathbb{Z} : x^3 - x^2 - 6x = 0\}$. Determine whether $S \subset T$ or $T \subset S$.

Solution: To see whether $S \subset T$ we must check whether $x \in S$ implies $x \in T$. Let $x \in S$. Then either $x = -2$ or $x = 3$. If $x = -2$ then $x^3 - x^2 - 6x = (-2)^3 - (-2)^2 - 6(-2) = 0$. Also, if $x = 3$ then $x^3 - x^2 - 6x = (3)^3 - (3)^2 - 6(3) = 0$. This verifies the syllogism "$x \in S$ implies $x \in T$." Therefore $S \subset T$.

The reverse inclusion fails, for $0 \in T$ but $0 \notin S$.   □

You Try It: How are $S = \{1, 2, 3, 5, 6, 8\}$ and $T = \{1, 3, 6\}$ related?

EXAMPLE 3.3
Let $S = \{x \in \mathbb{N} : x \geq 4\}$ and $T = \{x \in \mathbb{N} : x < 9\}$. Is it true that either $S \subset T$ or $T \subset S$?

Solution: Both inclusions are false. For $10 \in S$ but $10 \notin T$ and $2 \in T$ but $2 \notin S$.   □

Proposition 3.1 *Let S and T be sets. Then $S = T$ if and only if both $S \subset T$ and $T \subset S$.*

Proof: If $S = T$ then, by definition, S and T have precisely the same elements. In particular, this means that $x \in S$ implies $x \in T$ and also $x \in T$ implies $x \in S$. That is, $S \subset T$ and $T \subset S$.

Now suppose that both $S \subset T$ and $T \subset S$. Seeking a contradiction, suppose that $S \neq T$. Then either there is some element of S that is not an element of T or there is some element of T that is not an element of S. The first eventuality contradicts $S \subset T$ and the second eventuality contradicts $T \subset S$. We conclude that $S = T$. $\square$

Definition 3.3 We let $\emptyset$ denote the set that contains no elements. That is, $\forall x, x \notin \emptyset$. We call $\emptyset$ the *empty set*.

EXAMPLE 3.4

If S is any set then $\emptyset \subset S$. To see this, notice that the statement "if $x \in \emptyset$ then $x \in S$" *must* be true because the hypothesis $x \in \emptyset$ is false. (Check the truth table for "if-then" statements.) This verifies that $\emptyset \subset S$. $\square$

EXAMPLE 3.5

Let $S = \{x \in \mathbb{N} : x + 2 \geq 19 \text{ and } x < 3\}$. Then S is a sensible set. There are no internal contradictions in its definition. But $S = \emptyset$. There are no elements in S. $\square$

Definition 3.4 Let S and T be sets. We say that x is an element of $S \cap T$ if both $x \in S$ and $x \in T$. We say that x is an element of $S \cup T$ if either $x \in S$ or $x \in T$.

We call $S \cap T$ the *intersection* of the sets S and T. We call $S \cup T$ the *union* of the sets S and T.

EXAMPLE 3.6

Let $S = \{x \in \mathbb{N} : 2 < x < 9\}$ and $T = \{x \in \mathbb{N} : 5 \leq x < 14\}$. Then $S \cap T = \{x \in \mathbb{N} : 5 \leq x < 9\}$, for these are the points common to both sets. And $S \cup T = \{x \in \mathbb{N} : 2 < x < 14\}$, for these are the points that are either in S or in T or in both. $\square$

Remark 3.1 Observe that the use of "or" in the definition of set union justifies our decision to use the "inclusive 'or' " rather than the "exclusive 'or' " in mathematics. See also Proposition 3.2 below.

EXAMPLE 3.7

Let $S = \{x \in \mathbb{N} : 1 \leq x \leq 5\}$ and $T = \{x \in \mathbb{N} : 8 < x \leq 12\}$. Then $S \cap T = \emptyset$, for the sets S and T have no elements in common. On the other hand, $S \cup T = \{x \in \mathbb{N} : 1 \leq x \leq 5 \text{ or } 8 < x \leq 12\}$. $\square$

Definition 3.5 Let S and T be sets. We say that $x \in S \backslash T$ if both $x \in S$ and $x \notin T$. We call $S \backslash T$ the *set-theoretic difference* of S and T.

EXAMPLE 3.8

Let $S = \{x \in \mathbb{N} : 2 < x < 7\}$ and $T = \{x \in \mathbb{N} : 5 \le x < 10\}$. Then we see that $S \backslash T = \{x \in \mathbb{N} : 2 < x < 5\}$ and $T \backslash S = \{x \in \mathbb{N} : 7 \le x < 10\}$. □

Definition 3.6 Suppose that we are studying subsets of a fixed set X. If $S \subset X$ then we use the symbol ${}^cS$ to denote $X \backslash S$. In this context we sometimes refer to X as the *universal set*. We call ${}^cS$ the *complement* of S (in X).

EXAMPLE 3.9

Let $\mathbb{N}$ be the universal set. Let $S = \{x \in \mathbb{N} : 3 < x \le 20\}$. Then

$$ {}^cS = \{x \in \mathbb{N} : 1 \le x \le 3\} \cup \{x \in \mathbb{N} : 20 < x\} $$

□

The next proposition puts our use of the "inclusive or" into context.

Proposition 3.2 *Let X be the universal set and $S \subset X$, $T \subset X$. Then*

> **(1)** ${}^c(S \cup T) = {}^cS \cap {}^cT$
>
> **(2)** ${}^c(S \cap T) = {}^cS \cup {}^cT$

Proof: We shall present this proof in detail since it is a good exercise in understanding both our definitions and our method of proof.

We begin with the proof of Part **(1)**. It is often best to treat the proof of the equality of two sets as two separate proofs of containment. (This is why Proposition 3.1 is important.) That is what we now do.

Let $x \in {}^c(S \cup T)$. Then, by definition, $x \notin (S \cup T)$. Thus x is neither an element of S nor an element of T. So both $x \in {}^cS$ and $x \in {}^cT$. Hence $x \in {}^cS \cap {}^cT$. We conclude that ${}^c(S \cup T) \subset {}^cS \cap {}^cT$. Conversely, if $x \in {}^cS \cap {}^cT$ then $x \notin S$ and $x \notin T$. Therefore $x \notin (S \cup T)$. As a result, $x \in {}^c(S \cup T)$. Thus ${}^cS \cap {}^cT \subset {}^c(S \cup T)$. Summarizing, we have ${}^c(S \cup T) = {}^cS \cap {}^cT$.

The proof of Part **(2)** is similar, but we include it for practice. Let $x \in {}^c(S \cap T)$. Then, by definition, $x \notin (S \cap T)$. Thus x is not both an element of S and an element of T. So either $x \in {}^cS$ or $x \in {}^cT$. Hence $x \in {}^cS \cup {}^cT$. We conclude that ${}^c(S \cap T) \subset {}^cS \cup {}^cT$. Conversely, if $x \in {}^cS \cup {}^cT$ then either $x \notin S$ or $x \notin T$. Therefore $x \notin (S \cap T)$. As a result, $x \in {}^c(S \cap T)$. Thus ${}^cS \cup {}^cT \subset {}^c(S \cap T)$. Summarizing, we have ${}^c(S \cap T) = {}^cS \cup {}^cT$. □

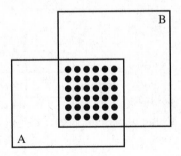

Figure 3.1   Venn diagram showing A ∩ B.

The two formulas in the last proposition are often referred to as *de Morgan's laws*. Compare them with de Morgan's laws for ∨ and ∧ at the end of Sec. 1.3.

3.3 Venn Diagrams

We sometimes use a *Venn diagram* to aid our understanding of set-theoretic relationships. In a Venn diagram, a set is represented as a region in the plane (for convenience, we use rectangles). The intersection $A \cap B$ of two sets A and B is the region common to the two domains (we have shaded that region with dots in Fig. 3.1):

The reader may wish to think about how to represent a *union* as a Venn diagram.

Now let A, B, and C be three sets. The Venn diagram in Fig. 3.2 makes it easy to see that $A \cap (B \cup C) = (A \cap B) \cup (A \cap C)$.

The Venn diagram in Fig. 3.3 illustrates the fact that

$$A \backslash (B \cup C) = (A \backslash B) \cap (A \backslash C)$$

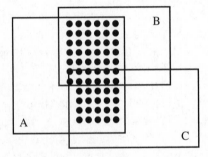

Figure 3.2   Venn diagram showing A ∩ (B ∪ C).

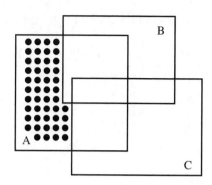

Figure 3.3   Venn diagram showing A\(B ∪ C).

A Venn diagram is not a proper substitute for a rigorous mathematical proof. However, it can go a long way toward guiding our intuition.

3.4 Further Ideas in Elementary Set Theory

Now we learn some new ways to combine sets.

Definition 3.7 Let S and T be sets. We define $S \times T$ to be the set of all ordered pairs (s, t) such that $s \in S$ and $t \in T$. The set $S \times T$ is called the *set-theoretic product* of S and T.

EXAMPLE 3.10
Let $S = \{1, 2, 3\}$ and $T = \{a, b\}$. Then

$$S \times T = \{(1, a), (1, b), (2, a), (2, b), (3, a), (3, b)\}$$

□

It is no coincidence that, in the last example, the set S has 3 elements, the set T has 2 elements, and the set $S \times T$ has $3 \times 2 = 6$ elements. In fact one can prove that if S has k elements and T has ℓ elements then $S \times T$ has $k \cdot \ell$ elements. Exercise 3.12 asks you to prove this assertion by induction on k.

Notice that $S \times T$ is a different set from $T \times S$. With S and T as in the last example,

$$T \times S = \{(a, 1), (b, 1), (a, 2), (b, 2), (a, 3), (b, 3)\}$$

The phrase "ordered pair" means that the pair $(a, 1)$, for example, is distinct from the pair $(1, a)$.

If S is a set then the *power set* of S is the set of all subsets of S. We denote the power set by $\mathcal{P}(S)$.

EXAMPLE 3.11
Let $S = \{1, 2, 3\}$. Then

$$\mathcal{P}(S) = \big\{\{1\}, \{2\}, \{3\}, \{1, 2\}, \{2, 3\}, \{1, 3\}, \{1, 2, 3\}, \emptyset\big\}$$

□

If the concept of power set is new to you, then you might have been surprised to see $\{1, 2, 3\}$ and $\emptyset$ as elements of the power set. But they are both subsets of S, and they must be listed.

Proposition 3.3 *Let $S = \{s_1, \ldots, s_k\}$ be a set. Then $\mathcal{P}(S)$ has 2^k elements.*

Proof: We prove the assertion by induction on k.

The inductive statement is

A set with k elements has power set with 2^k elements

$P(1)$ **is true.** In this case, $S = \{s_1\}$ and $\mathcal{P}(S) = \{\{s_1\}, \emptyset\}$. Notice that S has $k = 1$ element and $\mathcal{P}(S)$ has $2^k = 2$ elements.

$P(k) \Rightarrow P(k+1)$ **is true.** Assume that any set of the form $S = \{s_1, \ldots, s_k\}$ has power set with 2^k elements. Now let $T = \{t_1, \ldots, t_k, t_{k+1}\}$. Consider the subset $T' = \{t_1, \ldots, t_k\}$ of T. By the inductive hypothesis, the power set of T' has 2^k elements.

Now $\mathcal{P}(T)$ certainly contains $\mathcal{P}(T')$ (that is, every subset of T' is also a subset of T). But it also contains each of the sets that is obtained by adjoining the element t_{k+1} to each subset of T'. There are 2^k instances of each type of set. Thus the total number of subsets of T is

$$2^k + 2^k = 2^{k+1}$$

Notice that we have indeed counted all subsets of T, since any subset either contains t_{k+1} or it does not.

Thus, assuming the validity of our assertion for k, we have proved its validity for $k + 1$. That completes our induction and the proof of the proposition. □

We have seen that the operation of set-theoretic product corresponds to the arithmetic product of natural numbers. And now we have seen that the operation of taking the power set corresponds to exponentiation.

Exercises

1. Let $S = \{1, 2, 3, 4, 5\}$, $T = \{3, 4, 5, 7, 8, 9\}$, $U = \{1, 2, 3, 4, 9\}$, $V = \{2, 4, 6, 8\}$. Calculate each of the following:

   a. $(S \cup V) \cap U$

   b. $(S \cap T) \cup U$

2. Let S be any set and let $T = \emptyset$. What can you say about $S \times T$?

3. Prove the following formulas for arbitrary sets S, T, and U. [*Hint*: You may find Venn diagrams useful to guide your thinking, but a Venn diagram is *not* a proof.]

   a. $S \cap (T \cup U) = (S \cap T) \cup (S \cap U)$

   b. $S \cup (T \cap U) = (S \cup T) \cap (S \cup U)$

4. Draw Venn diagrams to illustrate parts (a) and (b) of Exercise 3.3.

5. Suppose that $A \subset B \subset C$. What is $A \backslash B$? What is $A \backslash C$? What is $A \cup B$?

6. Describe the set $\mathbb{Q} \backslash \mathbb{Z}$ in words. Describe $\mathbb{R} \backslash \mathbb{Q}$.

7. Describe $\mathbb{Q} \times \mathbb{R}$ in words. Describe $\mathbb{Q} \times \mathbb{Z}$.

8. Describe $(\mathbb{Q} \times \mathbb{R}) \backslash (\mathbb{Z} \times \mathbb{Q})$ in words.

9. Give an explicit description of the power set of $S = \{a, b, 1, 2\}$.

10. Let $S = \{a, b, c, d\}$, $T = \{1, 2, 3\}$, and $U = \{b, 2\}$. Which of the following statements is true?

    a. $\{a\} \in S$

    b. $1 \in T$

    c. $\{b, 2\} \in U$

11. Write out the power set of each set:

    a. $\{1, \emptyset, \{a, b\}\}$

    b. $\{\bullet, \triangle, \partial\}$

12. Prove using induction on k that if the set S has k elements and the set T has l elements then the set $S \times T$ has $k \cdot l$ elements.

CHAPTER 4

Functions and Relations

4.1 A Word About Number Systems

For the record, we make a note here of some of the number systems that will be used throughout this book. A more detailed and rigorous treatment of these mathematical constructs may be found in Chap. 5.

The most basic and rudimentary number system is the *natural numbers*. These are the counting numbers 1, 2, 3, ... There are infinitely many natural numbers. Of course the natural numbers are the number system that is most familiar to *everyone*. Everyone must count things—his/her possessions, or change from a purchase, or sports scores. Natural numbers are the key to all these processes. We denote the set of natural numbers with the symbol $\mathbb{N}$.

The next level of sophistication is represented by the *integers*. These are the whole numbers which are positive or negative or 0. Symbolically, the integers are given by

$$\ldots, -3, -2, -1, 0, 1, 2, 3, \ldots$$

From a mathematical point of view, the integers are more attractive than the natural numbers because they are closed under certain arithmetic operations—notably substraction. The expression $3 - 7$ makes good sense in the integers; it does *not* in the natural numbers. We denote the set of integers by $\mathbb{Z}$ (because "Z" is the first letter of the German word *Zahlen*, meaning *numbers*).

While the integers are closed under addition, subtraction, and multiplication, they are *not* closed under division. As an example, $5/7$ makes no sense in the integers. For this reason we create the number system known as the rational numbers. These are all fractions p/q, where p and q are integers and q is not equal to zero (because of course we are never allowed to divide by 0). The rational numbers form an attractive number system because they are closed under all four arithmetic operations. We denote the rational numbers by $\mathbb{Q}$ (standing for "quotient").

The most subtle and sophisticated number system, from our point of view, is the *real number system*. The real numbers consist of all decimal expansions, both terminating and nonterminating. All the rational numbers are also real numbers (and a rational number has a decimal expansion that is either terminating or repeating). But there are also decimal expansions that are both *nonterminating* and *nonrepeating*. These represent the *irrational numbers*—which are real numbers that are not rational. Most of modern science and engineering is done with the real number system. The real numbers are not only closed under the four basic arithmetic operations, but they are also closed under various limiting processes that are important for mathematical analysis. We denote the real number system by $\mathbb{R}$.

In closing, we shall briefly mention the *complex number system*. These are numbers of the form $x + iy$ where x and y are both real (and i denotes the square root of -1). The complex numbers have an addition operation and a multiplication/division operation—and the number system is closed under both of these. The complex numbers were invented to be a number system in which every polynomial equation has a root. But complex numbers have proved to be important in physics and engineering and partial differential equations. They are fundamental to modern mathematics and science. However, we shall see little of the complex numbers in the present book. The complex number system is denoted by $\mathbb{C}$.

It is worth noting that we have presented the number systems in order of sophistication. Each new number system was created because of some lack in the preceding number system. For instance, the integers were created because the natural numbers were not closed under subtraction. The rational numbers were created because the integers are not closed under division. And so forth.

Chapter 5 provides some detailed discussion of the number systems. In the remainder of the book we shall frequently use various number systems to illustrate important ideas from set theory or discrete mathematics.

4.2 Relations and Functions

Let S and T be sets. A *relation* on S and T is a subset of $S \times T$. If $\mathcal{R}$ is a relation then we write either $(s, t) \in \mathcal{R}$ or sometimes $s \mathcal{R} t$ to indicate that (s, t) is an element of the relation. We will also write $s \sim t$ when the relation being discussed is understood.

EXAMPLE 4.1

Let $S = \mathbb{N}$, the natural numbers (or positive, whole numbers); and let $T = \mathbb{R}$, the real numbers. Define a relation $\mathcal{R}$ by $(s, t) \in \mathcal{R}$ if $s < \sqrt{t} < s + 1$. For instance, $(2, 5) \in \mathcal{R}$ because $\sqrt{5}$ lies between 2 and 3. Also $(4, 17) \in \mathcal{R}$ because $\sqrt{17}$ lies between 4 and 5. However, $(5, 10)$ does not lie in $\mathcal{R}$ because $\sqrt{10}$ is *not* between 5 and $5 + 1 = 6$.   □

The *domain* of a relation $\mathcal{R}$ is the set of $s \in S$ such that there exists a $t \in T$ with $(s, t) \in \mathcal{R}$. The *image* of the relation is the set of $t \in T$ such that there exists an $s \in S$ with $(s, t) \in \mathcal{R}$. It is sometimes convenient to refer to the entire set T as the *range* of the relation $\mathcal{R}$. Some sources use the word "codomain" rather than "range". Clearly the range of a relation contains its image.

EXAMPLE 4.2

Let $S = \mathbb{N}$ and $T = \mathbb{N}$. Define a relation $\mathcal{R}$ on S and T by the condition $(s, t) \in \mathcal{R}$ if $s^2 < t$. Observe that, for any element $s \in \mathbb{N} = S$, the number $t = s^2 + 1$ satisfies $s^2 < t$. Therefore every $s \in S = \mathbb{N}$ is in the domain of the relation.

Now let us think about the image. The number $1 \in \mathbb{N} = T$ cannot be in the image since there is no element $s \in S = \mathbb{N}$ such that $s^2 < 1$. However, any element $t \in T$ that exceeds 1 satisfies $1^2 < t$. So $(1, t) \in \mathcal{R}$. Thus the image of $\mathcal{R}$ is the set $\{t \in \mathbb{N} : t \geq 2\}$.   □

EXAMPLE 4.3

Let $S = \mathbb{N}$ and $T = \mathbb{N}$. Define a relation $\mathcal{R}$ on S and T by the condition $(s, t) \in \mathcal{R}$ if $s^2 + t^2$ is itself a perfect square. Then, for instance, $(3, 4) \in \mathcal{R}$, $(4, 3) \in \mathcal{R}$, $(12, 5) \in \mathcal{R}$, and $(5, 12) \in \mathcal{R}$. The number 1 is not in the domain of $\mathcal{R}$ since there is no natural number t such that $1^2 + t^2$ is a perfect square (if there were, this would mean that there are two perfect squares that differ by 1, and that is not the case). The number 2 is not in the domain of $\mathcal{R}$ for a similar reason. Likewise, 1 and 2 are not in the image of $\mathcal{R}$.

In fact both the domain and image of $\mathcal{R}$ have infinitely many elements. This assertion will be explored in the exercises.   □

You Try It: Let S be the real numbers and T be the integers. Let us say that $s \in S$ is related to $t \in T$ if $s < t^3$. What are the domain, range, and image of this relation?

Many interesting relations arise for which S and T are the same set. Say that $S = T = A$. Then a relation on S and T is called simply a relation on A.

EXAMPLE 4.4

Let $\mathbb{Z}$ be the integers. Let us define a relation $\mathcal{R}$ on $\mathbb{Z}$ by the condition $(s, t) \in \mathcal{R}$ if $s - t$ is divisible by 2. It is easy to see that both the domain and the image of this relation is $\mathbb{Z}$ itself. It is also worth noting that if n is any integer then the set of all elements related to n is either (**a**) the set of all even integers (if n is even) or (**b**) the set of all odd integers (if n is odd). ☐

Notice that the last relation created a division of the domain (=image) into two disjoint sets: the even integers and the odd integers. This was a special instance of an important type of relation that we now define.

Definition 4.1 Let $\mathcal{R}$ be a relation on a set A. We say that $\mathcal{R}$ is an *equivalence relation* if the following properties hold:

  $\mathcal{R}$ **is reflexive:** If $x \in A$ then $(x, x) \in \mathcal{R}$;

  $\mathcal{R}$ **is symmetric:** If $(x, y) \in \mathcal{R}$ then $(y, x) \in \mathcal{R}$;

  $\mathcal{R}$ **is transitive:** If $(x, y) \in \mathcal{R}$ and $(y, z) \in \mathcal{R}$ then $(x, z) \in \mathcal{R}$.

Check for yourself that the relation described in Example 4.4 is in fact an equivalence relation. The most important property of equivalence relations is that which we indicated just before the definition and which we now enunciate formally:

Proposition 4.1 *Let $\mathcal{R}$ be an equivalence relation on a set A. If $x \in A$ then define*

$$E_x \equiv \{y \in A : (x, y) \in \mathcal{R}\}$$

We call the sets E_x the equivalence classes induced by the relation $\mathcal{R}$. If now s and t are any two elements of A then either $E_s \cap E_t = \emptyset$ or $E_s = E_t$.

In summary, the set A is the pairwise disjoint union of the equivalence classes induced by the equivalence relation $\mathcal{R}$.

Before we illustrate this proposition, let us discuss for a moment what it means. Clearly every element $a \in A$ is contained in some equivalence class, for a is contained in E_a itself. The proposition tells us that the set A is in fact the pairwise disjoint union of these equivalence classes. We say that the equivalence classes *partition* the set A.

For instance, in Example 4.4, the equivalence relation gives rise to two equivalence classes: the even integers $\mathcal{E}$ and the odd integers $\mathcal{O}$. Of course $\mathbb{Z} = \mathcal{E} \cup \mathcal{O}$

and $\mathcal{E} \cap \mathcal{O} = \emptyset$. We say that the equivalence relation *partitions* the universal set $\mathbb{Z}$ into two equivalence classes.

Notice that, in this example, if we pick any element $n \in \mathcal{E}$ then $E_n = \mathcal{E}$. Likewise, if we pick any element $m \in \mathcal{O}$, then $E_m = \mathcal{O}$.

The reason that an equivalence relation partitions its domain into pairwise disjoint equivalence classes is quite simple. For if E_{s_1} and E_{s_2} are two distinct equivalence classes that have nontrivial intersection, suppose that x lies in each of these sets. Then every element of E_{s_1} is related to x, and every element of E_{s_2} is related to x. By transitivity, it then follows that every element of E_{s_1} is related to every element of E_{s_2} (and vice versa!). So in fact E_{s_1} and E_{s_2} are one and the same equivalence class.

EXAMPLE 4.5
Let A be the set of all people in the United States. If $x, y \in A$ then let us say that $(x, y) \in \mathcal{R}$ if x and y have the same surname (that is, last name). Then $\mathcal{R}$ is an equivalence relation:

1. $\mathcal{R}$ is reflexive since any person x has the same surname as his/her self.
2. $\mathcal{R}$ is symmetric since if x has the same surname as y then y has the same surname as x.
3. $\mathcal{R}$ is transitive since if x has the same surname as y and y has the same surname as z then x has the same surname as z.

Thus $\mathcal{R}$ is an equivalence relation. The equivalence classes are all those people with surname Smith, all those people with surname Herkimer, and so forth.   □

EXAMPLE 4.6
Let S be the set of all residents of the United States. If $x, y \in S$ then let us say that x is related to y (that is, $x \sim y$) if x and y have at least one biological parent in common. It is easy to see that this relation is reflexive and symmetric. It is *not* transitive, as children of divorced parents know too well.

To be more explicit, John and Sally could have Bill as a child. Then John and Sally could get divorced, John could marry Bettie, and John and Bettie could have a child Bobby. Thus Bill and Bobby are "related" (under the relation described in the last paragraph). But John and Bettie could get divorced and then Bettie could marry Joe and have a child Rufus with Joe. So Bobby and Rufus are related.

We see that Bill and Bobby are related. And Bobby and Rufus are related. But certainly Bill and Rufus are *not* related, because they have no parent in common. So the relation is not transitive.

What this tells us (mathematically) is that the proliferation of divorce in our society does *not* lead to well-defined families.   □

4.3 Functions

In more elementary mathematics courses we define a function as follows: Let S and T be sets. A function f from S to T is a *rule* that assigns to each element of S a unique element of T.

This definition is problematic. The main difficulty is the use of the words "rule" and "assign." For instance, let $S = T = \mathbb{Z}$. Consider

$$f(x) = \begin{cases} x^2 & \text{if there is life as we know it on Mars} \\ 3x - 5 & \text{if there is not life as we know it on Mars} \end{cases}$$

Is this a function? Can what we see on the right be considered a rule? Do we have to wait until we have found life on Mars (or not!) before we can consider this a function?

More significantly in practice, thinking of a function as a rule is extremely limiting. The functions

$$f(x) = x^3 - 3x + 1$$

$$g(x) = \sin x$$

$$h(x) = \frac{\ln x}{x^2 + 4}$$

are inarguably given by rules. But open up your newspaper and look on the financial page at the graph of the Gross National Product. This is certainly the graph of a function, but what "rule" describes it?

Consider now the function

$$h(x) = \begin{cases} x^2 & \text{if } -\infty < x < -3 \\ 2x & \text{if } -3 \le x < 2 \\ x + 5 & \text{if } 2 \le x < \infty \end{cases}$$

For many years, until well into the twentieth century, mathematicians disagreed on whether this is a *bona fide* function. But it can certainly happen in practice that a physical process or a feedback mechanism can give data that is specified in a piecewise fashion. We need a more rigorous, and simultaneously a more flexible, means of understanding the concept of function.

It is best in advanced mathematics to have a way to think about functions that avoids subjective words like "rule" and "assign." This is the motivation for our next definition.

Definition 4.2 Let S and T be sets. A *function* f from S to T is a relation on S and T such that

(1) Every $s \in S$ is in the domain of f;

(2) If $(s, t) \in f$ and $(s, u) \in f$ then $t = u$.

Of course we refer to S and T as the domain and the range, respectively, of f.[1]
Condition (1) mandates that each element s of S is associated to *some* element of T.
Condition (2) mandates, in a formal manner, that each element s of S is associated
to *only one* member of T. Notice, however, that the definition neatly sidesteps the
notions of "assign" or "rule."

We shall frequently speak of the *image* of a given function f from S to T. This
just means the set that is the image of f when it is thought of as a relation. It is just
the set of elements $t \in T$ such that there is an $s \in S$ with $(s, t) \in f$.

EXAMPLE 4.7
Let $S = \{1, 2, 3\}$ and $T = \{a, b, c\}$. Set

$$f = \{(1, a), (2, a), (3, b)\}$$

This is a function, for it satisfies the properties set down in Definition 4.2. Given
the way that you are accustomed to writing functions in earlier courses, you might
find it helpful to view this function as

$$f(1) = a$$
$$f(2) = a$$
$$f(3) = b$$

Notice that each element $1, 2, 3$ of the domain is "assigned" to one and only
one element of the range. However, the definition of function allows the possibility
that two different elements of the domain be assigned to the same range element.
Observe that, for this function, the image is $\{a, b\}$ while the range is $\{a, b, c\}$. □

EXAMPLE 4.8
Let $S = \{1, 2, 3\}$ and $T = \{a, b, c, d, e\}$. Set

$$f = \{(1, b), (2, a), (3, c), (4, a), (5, b)\}$$

---

[1]We note, once again, that some sources use the word "codomain" instead of "range."

This is a function, for it satisfies the properties set down in Definition 4.2. Notice that each element of the domain S is used once and only once. However, not all elements of the range are used. According to the definition of function, this is allowed. Note that the *range* of f is $\{a, b, c, d, e\}$ while the *image* of f is $\{a, b, c\}$. □

Definition 4.3 Let f be a function with domain S and range T. We often write such a function as $f : S \to T$. We say that f is *one-to-one* or *injective* if whenever $(s, t) \in f$ and $(s', t) \in f$ then $s = s'$. We sometimes refer to such a mapping as an *injection*. We also refer to such a map as *univalent*.

Compare this new definition with Definition 4.2 of function. The new condition is similar to condition (2) for functions. But it is *not* the same. We are now mandating that no two domain elements be associated with the same range element.

EXAMPLE 4.9
Let $S = T = \mathbb{R}$ and let f be the set of all ordered pairs $\{(x, x^2) : x \in \mathbb{R}\}$. We may also write this function as

$$f : \mathbb{R} \to \mathbb{R}$$

$$x \mapsto x^2$$

or as $f(x) = x^2$.

It is easy to verify that f satisfies the definition of function. However, both of the ordered pairs $(-2, 4)$ and $(2, 4)$ are in f [in other words $f(-2) = 4 = f(2)$] so that f is *not* one-to-one. □

EXAMPLE 4.10
Let $S = T = \mathbb{R}$ and let f be the function $f(x) = x^3$. Then f is strictly increasing as x moves from left to right. In other words, if $s < t$ then $f(s) < f(t)$. Hence $f(s) \neq f(t)$. It follows that the function f *is* one-to-one. □

Definition 4.4 Let f be a function with domain S and range T. If for each $t \in T$ there is an $s \in S$ such that $f(s) = t$ then we say that f is *onto* or *surjective*. We sometimes refer to such a mapping as a *surjection*. Notice that a function is onto precisely when its image equals its range.

EXAMPLE 4.11
Let $f(x) = x^2$ be the function from Example 4.9. Recall from that example that $S = T = \mathbb{R}$. The point $t = -1 \in T$ has the property that there is no $s \in S$ such that $f(s) = t$. As a result, this function f is *not* onto. □

EXAMPLE 4.12
Let $S = \mathbb{R}, T = \{x \in \mathbb{R} : 1 \leq x < \infty\}$. Let $g : S \to T$ be given by $g(x) = x^2 + 1$.
Then for each $t \in T$ the number $s = +\sqrt{t - 1}$ makes sense and lies in S. Moreover,
$g(s) = t$. It follows that this function g is surjective. However, g is not injective
[since, for example, $g(-1) = g(1) = 2$].   □

4.4 Combining Functions

There are several elementary operations that allow us to combine functions in
useful ways. In this section, and from now on, we shall (whenever possible) write
our functions in the form

$$f(x) = \langle formula \rangle$$

for the sake of clarity. However we must keep in mind, and we shall frequently see,
that many functions *cannot* be expressed with an elegant formula.

Definition 4.5 Let f and g be functions with the same domain S and the same
range T. Assume that T is a set in which the indicated arithmetic operation (below)
makes sense. Then we define

(1) $(f + g)(x) = f(x) + g(x)$;
(2) $(f - g)(x) = f(x) - g(x)$;
(3) $(f \cdot g)(x) = f(x) \cdot g(x)$;
(4) $(f/g)(x) = f(x)/g(x)$ provided that $g(x) \neq 0$.

Notice that in each of **(1)**–**(4)** we are defining a *new function*—either $f + g$ or
$f - g$ or $f \cdot g$ or f/g—in terms of the component functions f and g. For practice,
we shall now express **(1)** in the language of ordered pairs. We ask you to do likewise
with **(2)**, **(3)**, **(4)** in Exercise 4.7.

Let us consider part **(1)** in detail. Now f is a collection of ordered pairs in $S \times T$
that satisfy the conditions for a function, and so is g. The function $f + g$ is given
by

$$f + g = \{(s, t + t') : (s, t) \in f, (s, t') \in g\}$$

Expressing the other combinations of f and g is quite similar, and you should
be sure to do the corresponding exercise.

EXAMPLE 4.13
Let $S = T = \mathbb{R}$. Define

$$f(x) = x^3 - x \quad \text{and} \quad g(x) = \sin(x^2)$$

Let us calculate $f + g$, $f - g$, $f \cdot g$, f/g.
  Now

$$(f + g)(x) = (x^3 - x) + \sin(x^2)$$

$$(f - g)(x) = (x^3 - x) - \sin(x^2)$$

$$(f \cdot g)(x) = (x^3 - x) \cdot [\sin(x^2)]$$

$$(f/g)(x) = \frac{x^3 - x}{\sin(x^2)} \quad \text{provided } x \neq \pm\sqrt{k\pi} \; k \in \{0, 1, 2, \ldots\} \qquad \square$$

  A more interesting, and more powerful, way to combine functions is through functional composition. Incidentally, in this discussion we will see the value of good mathematical notation.

Definition 4.6 Let $f : S \to T$ be a function and let $g : T \to U$ be a function. Then we define, for $s \in S$, the composite function

$$(g \circ f)(s) = g(f(s)) \tag{4.1}$$

We call $g \circ f$ the *composition* of the functions g and f.

Notice in this definition that the right-hand side of Eq. (4.1) always makes sense because of the way that we have specified the domain and range of the component functions f and g. In particular, we must have image $f \subset$ domain g in order for the composition to make sense—just because we are applying g to $f(s)$.

EXAMPLE 4.14
Let $f : \mathbb{R} \to \{x \in \mathbb{R} : x \geq 0\}$ be given by $f(x) = x^4 + x^2 + 6$ and $g : \{x \in \mathbb{R} : x \geq 0\} \to \mathbb{R}$ be given by $g(x) = \sqrt{x} - 4$. Notice that f and g fit the paradigm specified in the definition of composition of functions. Then

$$(g \circ f)(x) = g(f(x))$$

$$= g(x^4 + x^2 + 6)$$

$$= \sqrt{x^4 + x^2 + 6} - 4$$

Notice that $f \circ g$ also makes sense and is given by

$$(f \circ g)(x) = f(g(x))$$

$$= f(\sqrt{x} - 4)$$

$$= [\sqrt{x} - 4]^4 + [\sqrt{x} - 4]^2 + 6$$

It is important to understand that $f \circ g$ and $g \circ f$, when both make sense, will generally be different.   □

It is a good exercise in the ideas of this chapter to express the notion of functional composition in the language of ordered pairs. Thus let $f : S \to T$ be a function and $g : T \to U$ be a function. Then f is a subset of $S \times T$ and g is a subset of $T \times U$, both satisfying the two standard conditions for functions. Now $g \circ f$ is a set of ordered pairs specified by

$$g \circ f =$$

$$\{(s, u) : s \in S, u \in U, \text{ and } \exists t \in T \text{ such that } (s, t) \in f \text{ and } (t, u) \in g\}$$

Take a moment to verify that this equation is consistent with the definition of functional composition that we gave earlier.

EXAMPLE 4.15

Let $f : \mathbb{R} \to [-1, 1]$ be given by $f(x) = \sin x^5$ and let $g : \{x \in \mathbb{R} : x \geq 1\} \to \mathbb{R}$ be given by $\sqrt[4]{x - 1}$. We cannot consider $g \circ f$ because the range of f (namely, the set $[-1, 1]$) does not lie in the domain of g. However, $f \circ g$ *does* make sense because the range of g lies in the domain of f. And

$$(f \circ g)(x) = \sin[(x - 1)^{5/4}] \qquad \text{for } x \geq 1$$

                                                                                      □

Definition 4.7 Let S and T be sets. Let $f : S \to T$ and $g : T \to S$. We say that f and g are *mutually inverse* provided that both $(f \circ g)(t) = t$ for all $t \in T$ and $(g \circ f)(s) = s$ for all $s \in S$. We write $g = f^{-1}$ or $f = g^{-1}$. We refer to the functions f and g as *invertible*; we call g the *inverse* of f and f the *inverse* of g.

EXAMPLE 4.16

Let $f : \mathbb{R} \to \mathbb{R}$ be given by $f(s) = s^3 - 1$ and $g : \mathbb{R} \to \mathbb{R}$ be given by $g(t) = \sqrt[3]{t+1}$. Then

$$(f \circ g)(t) = \left[\sqrt[3]{t+1}\right]^3 - 1$$
$$= (t+1) - 1$$
$$= t$$

for all t and

$$(g \circ f)(s) = \sqrt[3]{(s^3 - 1) + 1}$$
$$= \sqrt[3]{s^3}$$
$$= s$$

for all s. Thus $g = f^{-1}$ (or $f = g^{-1}$). □

EXAMPLE 4.17

Let $g : [1, +\infty) \to [8, +\infty)$ be given by $g(x) = x^2 + 3x + 4$. Then $g(1) = 8$, and g is strictly increasing without bound. So g is one-to-one and onto. We conclude that g has an inverse. One may use the quadratic formula to solve for that inverse, and find that $g^{-1}(x) = (-3 + \sqrt{-7 + 4x})/2$. □

The idea of inverse function lends itself particularly well to the notation of ordered pairs. For $f : S \to T$ is inverse to $g : T \to S$ (and vice versa) provided that for every ordered pair $(s, t) \in f$ there is an ordered pair $(t, s) \in g$ and conversely.

Not every function has an inverse. For instance, let $f : S \to T$. Suppose that $f(s) = t$ and also that $f(s') = t$ with $s \neq s'$ (in other words, suppose that f is not one-to-one). If $g : T \to S$ then $g(f(s)) = g(t) = g(f(s'))$ so it cannot be that both $g(f(s)) = s$ and $g(f(s')) = s'$. In other words, f cannot have an inverse. We conclude that a function that *does* have an inverse must be one-to-one.

On the other hand, suppose that $t \in T$ has the property that there is no $s \in S$ with $f(s) = t$ (in other words, suppose that f is not onto). Then, in particular, it could not be that $f(g(t)) = t$ for any function $g : T \to S$. So f could not be invertible. We conclude that a function that *does* have an inverse must be onto.

EXAMPLE 4.18

Let $f : \mathbb{R} \to \{x \in \mathbb{R} : x \geq 0\}$ be given by $f(x) = x^2$. Then f is onto, but f is not one-to-one. It follows that f cannot have an inverse. And indeed it does not, for any attempt to produce an inverse function runs into the ambiguity that every positive number has two square roots.

Let $f : \{x \in \mathbb{R} : x \geq 0\} \rightarrow \mathbb{R}$ be given by $f(x) = x^3$. Then f is one-to-one but f is not onto. There certainly is a function $g : \mathbb{R} \rightarrow \{x \in \mathbb{R} : x \geq 0\}$ such that $(g \circ f)(x) = x$ for all $x \in \{x \in \mathbb{R} : x \geq 0\}$ [namely $g(x) = \sqrt[3]{|x|}$]. But there is no function $h : \mathbb{R} \rightarrow \{x \in \mathbb{R} : x \geq 0\}$ such that $(f \circ h)(x) = x$ for all x. □

We have established that *if* $f : S \rightarrow T$ has an inverse *then* f must be one-to-one and onto. The converse is true too, and we leave the details for you to verify. A function $f : S \rightarrow T$ that is one-to-one and onto (and therefore invertible) is sometimes called a *set-theoretic isomorphism* or a *bijection*. It is also common to use the terminology *one-to-one correspondence*.

EXAMPLE 4.19
The function $f : \mathbb{R} \rightarrow \mathbb{R}$ that is given by $f(x) = x^3$ is a bijection. You should check the details of this assertion for yourself. The inverse of this function f is the function $g : \mathbb{R} \rightarrow \mathbb{R}$ given by $g(x) = x^{1/3}$. □

We leave it as an exercise for you to verify that the composition of two bijections (when the composition makes sense) is a bijection.

4.5 Types of Functions

The most elementary and easiest understood type of function is the polynomial function. A *polynomial* has the form

$$p(x) = a_0 + a_1 x + a_2 x^2 + a_3 x^3 + \cdots + a_k x^k$$

We call a_0 the *constant coefficient*, a_1 the *linear coefficient*, a_2 the *quadratic coefficient*, a_j the *jth-degree coefficient*, and a_k the *top-order coefficient*.

A polynomial is easy to understand because we can calculate its value at any point x by simply plugging in x and then multiplying and adding. A computer can calculate values of a polynomial very rapidly and easily.

Another important type of function is the *exponential function*. For example, $g(x) = 2^x$ is an exponential function. This function is easy to calculate when x is an integer; for example, $f(4) = 2^4 = 2 \cdot 2 \cdot 2 \cdot 2 = 16$. It is more difficult to calculate when x is a noninteger; some approximation procedure would probably be needed (this is how your calculator effects the computation).

You probably know that, when you put money in the bank, and it earns interest, then it grows according to a rule given by an exponential function. For example, suppose that you put $1000 in the bank and it earns interest annually at the rate of 5%. Then we have

Year	Amount
After the first year	$1000 \cdot 1.05$
After the second year	$1000 \cdot (1.05)^2$
After the third year	$1000 \cdot (1.05)^3$
...	
After the k th year	$1000 \cdot (1.05)^k$

The most important fact about an exponential function is that it will grow faster than a polynomial function. We can illustrate this fact simply by doing some calculations with a calculator:

Variable x	$f(x) = x^2$	$g(x) = 2^x$
1	1	2
2	4	4
3	9	8
4	16	16
10	100	1024
50	2500	1.125×10^{15}
100	10000	1.26×10^{30}

It can be proved in general that if p is *any polynomial* and f is *any exponential function with positive exponent* then, for x large enough (and positive), $f(x) > p(x)$. In fact we can say more: For x large enough, $f(x) > 100 \cdot p(x)$. Or, for x even larger, $f(x) > 1000 \cdot p(x)$. As we see from the last table, the growth of exponential functions is dramatic.

The flip side of this last information is that logarithmic functions—a third important type of function—grow very slowly (in fact slower than any polynomial). A logarithmic function is simply the inverse of an exponential function. For example, $h(x) = \log_2 x$ is the inverse of $g(x) = 2^x$. We illustrate the idea with another table:

Variable x	$f(x) = x^2$	$g(x) = \log_2 x$
1	1	0
2	4	1
4	16	2
16	256	4
32	1024	5
128	16384	7

It can be proved in general that if p is *any polynomial with positive coefficients* and g is *any logarithmic function* then, for x large enough (and positive), $g(x) < p(x)$. In fact we can say more: For x large enough, $g(x) < (1/100) \cdot p(x)$. Or, for x even larger, $g(x) < (1/1000) \cdot p(x)$. As we see from the last table, the growth of logarithmic functions is dramatically slow.

Many physical phenomena—such as entropy—are best described using logarithmic functions. The intensity of an earthquake is measured using a logarithmic (the *Richter*) scale.

Exercises

1. Consider the relation on $\mathbb{Z}$ defined by $(m, n) \in \mathcal{R}$ if $m + n$ is even. Prove that this is an equivalence relation. What are the equivalence classes?

2. Consider the relation on $\mathbb{Z} \times (\mathbb{Z} \setminus \{0\})$ defined by $(m, n)\mathcal{R}(m', n')$ provided that $m \cdot n' = m' \cdot n$. Prove that this is an equivalence relation. Can you describe the equivalence classes?

3. Consider the relation defined on the cartesian plane by $(x, y)\mathcal{R}(x', y')$ if $y = y'$. Prove that this is an equivalence relation. Can you describe the equivalence classes? Can you pick a representative (that is, an element) of each equivalence class that will help to exhibit what the equivalence relation is?

4. Let S be the set of all living people. Let $x, y \in S$. Say that x is related to y if x and y are siblings or the same person, that is, x and y have *both* parents the same. Prove that this is an equivalence relation. What are the equivalence classes?

5. Let $S = \{a, b, c, d\}$ and $T = \{1, 2, 3, 4, 5, 6, 7\}$. Which of the following relations on $S \times T$ is a function? Why?

   a. $\{(a, 4), (d, 3), (c, 3), (b, 2)\}$
   b. $\{(a, 5), (c, 4), (d, 3)\}$

6. Which of the following functions is one-to-one? Which is onto?

   a. $f : \mathbb{N} \to \mathbb{N}$ $f(m) = m + 2$
   b. $g : \mathbb{Z} \to \mathbb{Z}$ $g(m) = 2m^2 - 7$

7. Express parts (2), (3), (4) of Definition 4.5 using the language of ordered pairs. Imitate our discussion of part (1) in the text.

8. Find the domain and image of each of these relations:

   a. $\{(x, y) \in \mathbb{R} \times \mathbb{R} : x = \sqrt{y + 3}\}$

   b. $\{(\alpha, \beta) : \alpha$ is a person, β is a person, and α is the father of $\beta\}$

9. Use your calculator (or computer) to help you determine at what value of x the function $f(x) = 3^{x^2 + 5x + 7}$ passes the function $g(x) = 1000$ $x^{10} + 100x^5 + 10x + 100000$ in size.

10. Discuss why there are infinitely many triples of natural number k, l, m such that $k^2 + l^2 = m^2$.

CHAPTER 5

Number Systems

5.1 Preliminary Remarks

The purpose of this chapter is to illustrate the general principle of *modeling*. An axiomatic theory has no meaning, indeed no essential validity, unless there exists a model consisting of a collection of (mathematical) objects that actually satisfy the properties specified in the axioms. And constructing a model of an axiomatic system is insurance that the system will never lead to a contradiction (subject to the usual cautions about relative consistency).

Thus our purpose is to actually construct the natural numbers, the integers, the rational numbers, the real numbers, the complex numbers, and so on. Again, the point is that there is no logical validity in saying "it seems to me that there ought to be some negative numbers floating around somewhere" or "it seems to me that the number -1 should have a square root." One must *construct number systems in which this is so*.

One misleading feature of the presentation in this chapter, or in any book constructing the number systems from first principles, is that the more sophisticated number systems appear to be easier to construct than the simpler ones. This is partly because of experience: after we've constructed four number systems then the fifth one follows familiar patterns. But more to the point is that any construction of the natural numbers, for instance, must confront fundamental issues of logic (such as well-ordering and induction). As a result, some nasty issues will come up in Sec. 5.2. The later number systems are founded on the earlier ones, so their presentation will be more fluid and more natural.

These nasty issues must be considered a part of the firmament. There is no simple way to deal with the basic problems connected with the natural number system. Every student of mathematics should be exposed to these issues at least once; then he/she can get on with the rest of the chapter, and with issues that are more directly related to the everyday use of mathematics.

5.2 The Natural Number System

Giuseppe Peano's axioms for the natural numbers are as follows. In this discussion, we will follow tradition and use the notation $'$ to denote the "successor" of a natural number. For instance, the successor of 2 is $2'$. Intuitively, the successor of n is the number $n + 1$. However, addition is something that comes later; so we formulate the basic properties of the natural numbers in terms of the successor function.

5.2.1 PEANO'S AXIOMS FOR THE NATURAL NUMBERS

P1 $1 \in \mathbb{N}$.

P2 If $n \in \mathbb{N}$ then $n' \in \mathbb{N}$.

P3 There is no natural number n such that $n' = 1$.

P4 If m and n are natural numbers and if $m' = n'$ then $m = n$.

P5 Let P be a property. If
  1. $P(1)$ is true;
  2. $P(k) \Rightarrow P(k')$ for every $k \in \mathbb{N}$
  then $P(n)$ is true for every $n \in \mathbb{N}$.

As Suppes says in [SUP, pp. 121 ff.], these axioms for the natural numbers are almost universally accepted (although E. Nelson [NEL], among others, has found it useful to explore how to develop arithmetic without Axiom **P5**). They have evolved into their present form so that the natural numbers will satisfy those properties that are generally recognized as desirable. Let us briefly mention what each of the axioms signifies:

P1 asserts that $\mathbb{N}$ contains a distinguished element that we denote by 1.

P2 asserts that each element of $\mathbb{N}$ has a successor.

P3 asserts that 1 is not the successor of any natural number; in other words, 1 is in a sense the "first" element of $\mathbb{N}$.

P4 asserts that if two natural numbers have the same successor then they are in fact the same natural number.

P5 asserts that the method of induction is valid.

Some obvious, and heuristically appealing, properties of the natural numbers can be derived rather directly from Peano's axioms. Here is an example:

Proposition 5.1 *Let n be a natural number other than 1. Then $n = m'$ for some natural number m' for some natural number m.*

Remark 5.1 This proposition makes the intuitively evident assertion that every natural number, except 1, is a successor. Another way of saying this is that every natural number except 1 has a predecessor. This claim is clearly not an explicit part of any of the five axioms. In fact the only axiom that has a hope of implying the proposition is the inductive axiom, as we shall now see.

Proof of Proposition 5.1:   Let $P(n)$ be the statement "either $n = 1$ or $n = m'$ for some natural number m.

Clearly $P(n)$ is true when $n = 1$.

Now suppose that the statement $P(n)$ has been established for some natural number n. We wish to establish it for n'. But n' is, by definition, a successor. So the statement is true.

This completes our induction.   □

This proof is misleading in its simplicity. The proof consists of little more than *interpreting* Axiom **P5**. Some other desirable properties of the natural numbers are much more difficult to achieve directly from the axioms. As an instance, to prove that there is no natural number lying between k and k' is complicated. Indeed, the entire concept of ordering is extremely tricky. And the single most important property of the natural numbers (one that is essentially equivalent to the induction axiom, as we shall later see) is that the natural numbers are well ordered in a canonical fashion. So we must find an efficient method for establishing the properties of the natural numbers that are connected with order.

It is generally agreed (see [SUP, p. 121 ff.]) that the best way to develop further properties of the natural numbers is to treat a specific model. Even that approach is nontrivial; so we shall only briefly sketch the construction of a model and further sketch its order properties so that we may discuss well ordering. The approach that we take is by way of the so-called *finite ordinal arithmetic*.

Let us define a model for the natural numbers as follows:

$$1 = \{\emptyset\}$$

$$2 = 1 \cup \{1\} = \{\emptyset, \{\emptyset\}\}$$

$$3 = 2 \cup \{2\} = \{\emptyset, \{\emptyset\}, \{\emptyset, \{\emptyset\}\}\}$$

and, in general,

$$k' = k \cup \{k\}$$

It is straightforward to verify Peano's axioms for this model of the natural numbers. Let us first notice that, in this model, the successor n' of a natural number n is given by $n' = n \cup \{n\}$. Now let us sketch the verification of the axioms.

P1 is clear by construction, and so is **P2**.

P3 is an amusing exercise in logic: if $m' = 1$ then there is a set A such that $A \cup \{A\} = 1$ or $A \cup \{A\} = \{\emptyset\}$. In particular, $x \in A \cup \{A\}$ implies $x \in \{\emptyset\}$. Since $A \in A \cup \{A\}$, it follows that $A = \emptyset$. But $\emptyset$ is not a natural number. So 1 is not the successor of a natural number.

For **P4**, it is convenient to invoke the concept of ordering in our model of the natural numbers. Say that $m < n$ if $m \in n$. Clearly, if $m, n \in \mathbb{N}$ and $m \neq n$ then either $m \in n$ or $n \in m$ but not both. Thus we have the usual trichotomy of a strict, simple order. Now suppose that m, n are natural numbers and that $m' = n'$. If $m < n$ then $m \in n$ so $m' \in n'$ and $m' < n'$. That is false. Likewise we cannot have $n < m$. Thus it must be that $m = n$.

We shall discuss **P5** a bit later.

Next we turn to well-ordering. We assert that our model of the natural numbers, with the ordering defined in our discussion of **P4**, has the property that if $\emptyset \neq S \subset \mathbb{N}$ then there is an element $s \in S$ such that $s < t$ for every $s \neq t \in S$. It is clear from the trichotomy that the least element s, if it exists, is unique. We proceed in several steps:

Fix a natural number $m > 1$ and restrict attention to $\mathcal{Q}(m)$, which is the set of natural numbers that are less than m.

Proposition 5.2 *The set $\mathcal{Q}(m)$ has finitely many elements.*

Proof: A natural number k is less than m if and only if $k \in m$. But, by construction, m has only finitely many elements.   □

Remark 5.2 As an exercise, you may wish to attempt to prove this last proposition directly from the Peano axioms.

Proposition 5.3 For each m, the set $\mathcal{Q}(m)$ is well ordered.

Proof: The proof is by induction on m.

When $m = 1$ there is nothing to prove.

Assume that the assertion has been proved for $m = k$. Now let U be a subset of $\mathcal{Q}(k')$. There are now three possibilities:

1. If in fact $U \subset \mathcal{Q}(k)$ then U has a least element by the inductive hypothesis.
2. If $U = \{k\}$ then U has but one element and that element, namely k, is the least element that we seek.
3. The last possibility is that U contains k and some other natural numbers as well. But then $U \setminus \{k\} \subset \mathcal{Q}(k)$. Hence $U \setminus \{k\}$ has a least element s by the inductive hypothesis. Since s is automatically less than k, it follows that s is a least element for the entire set U. $\square$

Now we have all our tools in place and we can prove the full result:

Theorem 5.1 *The natural numbers $\mathbb{N}$ are well ordered.*

Proof: Let $\emptyset \neq S \subset \mathbb{N}$. Select an element $m \in S$. There are now two possibilities:

1. If $\mathcal{Q}(m) \cap S = \emptyset$ then m is the least element of S that we seek.
2. If $T = \mathcal{Q}(m) \cap S \neq \emptyset$ then notice that if $x \in T$ and $y \in S \setminus T$ then $x < y$. So it suffices for us to find a least element of T. But such an element exists by applying the preceding proposition to $U = \mathcal{Q}(m) \cap S$.

The proof is complete. $\square$

We next observe that, in a certain sense, the well-ordering property implies the induction property (Axiom **P5**). By this we mean the following: Do not consider any model of the natural numbers, but just consider any number system X satisfying **P1–P4**. Assume that every element of X except 1 has a predecessor, and in addition assume that this number system is well ordered.

Now let P be a property. Assume that $P(1)$ is true and assume the syllogism $P(k) \Rightarrow P(k')$. We claim that $P(n)$ is true for every n. Suppose not. Then the set

$$S = \{m \in X : P(m) \text{ is false}\}$$

is nonempty. Let q be the least element of S; this number is guaranteed to exist by the well-ordering property. The number q cannot be 1, for we assumed that $P(1)$ is true. But if $q > 1$ then q has a predecessor r. Since q is the least element of S then it cannot be that $r \in S$. Thus $P(r)$ must be true. But then, by our hypothesis, $P(r')$ must be true. However, $r' = q$. So $P(q)$ is true. That is a contradiction.

The only possible conclusion is that S is empty. So $P(k)$ is true for every k.

These remarks about well ordering implying **P5** are not satisfactory from our point of view, because *we used the inductive property* to establish that every natural number other than 1 has a predecessor. But it is important for you to understand that any development of the natural numbers will result in induction and well ordering being closely linked.

If you review your calculus book or other elementary texts, you will find that both induction and well ordering are occasionally used. But in every instance the author will say "These properties of the natural numbers are intuitively clear; trust me." Now you can begin to understand why an elementary textbook author must make that choice. The truth about these topics is inexorably linked to the very foundations of mathematics, and is therefore both subtle and complicated.

The remaining big idea connected with the natural numbers is addition. It can be proven that a satisfactory theory of addition *cannot* be developed from **P1–P5**. Instead, it is customary to adjoin two new axioms to our theory:

P6 If x is a natural number then $x + 1 = x'$;

P7 For any natural numbers x, y we have

$$x + y' = (x + y)'$$

To illustrate these ideas, let us close the section by proving that $2 + 2 = 4$. (Judge for yourself whether the proof is as obvious as $2 + 2 = 4$!) In this argument we use the definitions $2 = 1', 3 = 2', 4 = 3'$.

$$
\begin{aligned}
2 + 2 &= 2 + 1' && \text{by definition of 2} \\
&= (2 + 1)' && \text{by **P7**} \\
&= (2')' && \text{by **P6**} \\
&= 3' && \text{by definition of 3} \\
&= 4 && \text{by definition of 4}
\end{aligned}
$$

In fact it is even trickier to get multiplication to work in the natural number system. Almost the only viable method is to add even more axioms that control this binary operation. We can say no more about the matter here, but refer to [SUP].

It is not difficult to see that, with enough patience (or with induction) one could establish all the basic laws of arithmetic. Of course this would not be a fruitful use of our time. The celebrated work [WHB] treats this matter in complete detail.

In the succeeding sections of the present book, we shall take the basic laws of arithmetic on the natural numbers as given. We understand that our treatment of the natural numbers is incomplete. We have touched on some topics, and indicated some constructions. But, when it comes right down to it, we are taking the natural numbers on faith. All of our future number systems (the integers, the rational numbers, the reals, the complexes, the quaternions) will be constructed rigorously. The somewhat bewildering situation before us is that the more complicated numbers systems are easier to construct.

5.3 The Integers

Now we will apply the notion of an equivalence class to *construct* the integers (both positive and negative). There is an important point of knowledge to be noted here. In view of Sec. 5.2, we may take the natural numbers as given. The natural numbers are universally accepted, and we have indicated how they may be constructed in a formal manner. However, the number zero and the negative numbers are a different matter. It was not until the fifteenth century that the concepts of zero and negative numbers started to take hold—for they do not correspond to explicit collections of objects (five fingers or ten shoes) but rather to *concepts* (zero books is the lack of books; minus four pens means that we owe someone four pens). After some practice we get used to negative numbers, but explaining in words what they mean is always a bit clumsy.

In fact it is sobering to realize that the Italian mathematicians of the fifteenth and sixteenth centuries referred to negative numbers—in their *formal writings*—as "fictitious" or "absurd." Mathematics is, in part, a subject that we must get used to. It took several hundred years for mankind to get used to negative numbers.

It is much more satisfying, from the point of view of logic, to *construct* the integers from what we already have, that is, from the natural numbers. We proceed as follows. Let $A = \mathbb{N} \times \mathbb{N}$, the set of ordered pairs of natural numbers. We define a relation $\mathcal{R}$ on A as follows:

$$(a, b) \text{ is related to } (a^*, b^*) \text{ if } a + b^* = a^* + b$$

Theorem 5.2 *The relation $\mathcal{R}$ is an equivalence relation.*

Proof: That (a, b) is related to (a, b) follows from the trivial identity $a + b = a + b$. Hence $\mathcal{R}$ is reflexive. Second, if (a, b) is related to (a^*, b^*) then $a + b^* = a^* + b$ hence $a^* + b = a + b^*$ (just reverse the equality) hence (a^*, b^*) is related to (a, b). So $\mathcal{R}$ is symmetric.

Finally, if (a, b) is related to (a^*, b^*) and (a^*, b^*) is related to (a^{**}, b^{**}) then we have

$$a + b^* = a^* + b \quad \text{and} \quad a^* + b^{**} = a^{**} + b^*$$

Adding these equations gives

$$(a + b^*) + (a^* + b^{**}) = (a^* + b) + (a^{**} + b^*)$$

Cancelling a^* and b^* from each side finally yields

$$a + b^{**} = a^{**} + b$$

Thus (a, b) is related to (a^{**}, b^{**}). Therefore $\mathcal{R}$ is transitive. We conclude that $\mathcal{R}$ is an equivalence relation. □

Remark 5.3 We cheated a bit in the proof of Theorem 5.2. Since we do not yet have negative numbers, we therefore have not justified the process of "cancelling" that we used. The most rudimentary form of cancellation is Axiom **P4** of the natural numbers. Suggest a way to use induction, together with Axiom **P4**, to prove that if a, b, c are natural numbers and if $a + b = c + b$ then $a = c$.

Now our job is to understand the equivalence classes which are induced by $\mathcal{R}$. Let $(a, b) \in A = \mathbb{N} \times \mathbb{N}$ and let $[(a, b)]$ be the corresponding equivalence class. If $b > a$ then we will denote this equivalence class by the integer $b - a$. For instance, the equivalence class $[(2, 7)]$ will be denoted by 5. Notice that if $(a^*, b^*) \in [(a, b)]$ then $a + b^* = a^* + b$ hence $b^* - a^* = b - a$ *as long as* $b > a$. Therefore the numeral that we choose to represent our equivalence class is *independent of which element of the equivalence class is used to compute it.*

If $(a, b) \in A$ and $b = a$ then we let the symbol 0 denote the equivalence class $[(a, b)]$. Notice that if (a^*, b^*) is any other element of this particular $[(a, b)]$ then it must be that $a + b^* = a^* + b$ hence $b^* = a^*$; therefore this definition is unambiguous.

If $(a, b) \in A$ and $a > b$ then we will denote the equivalence class $[(a, b)]$ by the symbol $-(a - b)$. For instance, we will denote the equivalence class $[(7, 5)]$ by the symbol -2. Once again, if (a^*, b^*) is related to (a, b) then the equation $a + b^* = a^* + b$ guarantees that our choice of symbol to represent $[(a, b)]$ is unambiguous.

Thus we have given our equivalence classes names, and these names *look just like* the names that we give to integers: there are positive integers, and negative ones, and zero. But we want to see that these objects *behave* like integers. (As you read on, use the informal mnemonic that the equivalence class $[(a, b)]$ stands for the integer $b - a$.)

First, do these new objects that we have constructed *add* correctly? Well, let $A = [(a, b)]$ and $C = [(c, d)]$ be two equivalence classes. *Define* their sum to be $A + C = [(a + c, b + d)]$. We must check that this is unambiguous. If $(\widetilde{a}, \widetilde{b})$ is related to (a, b) and $(\widetilde{c}, \widetilde{d})$ is related to (c, d) then of course we know that

$$a + \widetilde{b} = \widetilde{a} + b$$

and

$$c + \widetilde{d} = \widetilde{c} + d$$

Adding these two equations gives

$$(a + c) + (\widetilde{b} + \widetilde{d}) = (\widetilde{a} + \widetilde{c}) + (b + d)$$

hence $(a + c, b + d)$ is related to $(\widetilde{a} + \widetilde{c}, \widetilde{b} + \widetilde{d})$. Thus adding two of our equivalence classes gives another equivalence class, as it should. We say that addition of integers is *well defined*.

This point is so significant that it bears repeating. Each integer is an equivalence class—that is, a *set*. If we are going to add two integers m and n by choosing an element from the set m and another element from the set n, then the operation that we define had better be independent of the choice of elements. This is another way of saying that we want the sum of two equivalence classes to be another equivalence class. We call this the concept of "well definedness."

EXAMPLE 5.1

To add 5 and 3 we first note that 5 is the equivalence class $[(2, 7)]$ and 3 is the equivalence class $[(2, 5)]$. We add them componentwise and find that the sum is $[(2 + 2, 7 + 5)] = [(4, 12)]$. Which equivalence class is this answer? Looking back at our prescription for giving names to the equivalence classes, we see that this is the equivalence class that we called $12 - 4$ or 8. So we have rediscovered the fact that $5 + 3 = 8$.

Now let us add 4 and -9. The first of these is the equivalence class $[(3, 7)]$ and the second is the equivalence class $[(13, 4)]$. The sum is therefore $[(16, 11)]$, and this is the equivalence class that we call $-(16 - 11)$ or -5. That is the answer that we would expect when we add 4 to -9.

Next, we add -12 and -5. Previous experience causes us to expect the answer to be -17. Now -12 is the equivalence class $[(19, 7)]$ and -5 is the equivalence class $[(7, 2)]$. The sum is $[(26, 9)]$, which is the equivalence class that we call -17.

Finally, we can see in practice that our method of addition is unambiguous. Let us redo the second example using $[(6, 10)]$ as the equivalence class denoted by 4 and $[(15, 6)]$ as the equivalence class denoted by -9. Then the sum is $[(21, 16)]$, and this is still the equivalence class -5, as it should be.   □

Remark 5.4 What is the point of this section? Everyone knows about negative numbers, so why go through this abstract construction? The reason is that until one sees this construction, negative numbers are just imaginary objects—placeholders if you will—which are a useful notation but which do not exist. Now they **do exist**. They are a collection of equivalence classes of pairs of natural numbers. This collection is equipped with certain arithmetic operations, such as addition, subtraction, and multiplication. We now discuss these last two.

If $A = [(a, b)]$ and $C = [(c, d)]$ are integers, then we define their *difference* to be the equivalence class $[(a + d, b + c)]$; we denote this difference by $A - C$.

(Note that we may not use subtraction of natural numbers in our definition of subtraction of integers; subtraction of natural numbers is not, in general, defined.) The unambiguity (or well definedness) of this definition is treated in the exercises.

EXAMPLE 5.2
We calculate $8 - 14$. Now $8 = [(1, 9)]$ and $14 = [(3, 17)]$. Therefore

$$8 - 14 = [(1 + 17, 9 + 3)] = [(18, 12)] = -6$$

as expected.

As a second example, we compute $(-4) - (-8)$. Now

$$-4 - (-8) = [(6, 2)] - [(13, 5)] = [(6 + 5, 2 + 13)] = [(11, 15)] = 4 \quad \square$$

Remark 5.5 When we first learn that $(-4) - (-8) = (-4) + 8 = 4$, the explanation is a bit mysterious: why is "minus a minus equal to a plus"? Now there is no longer any mystery: this property follows from *our construction* of the number system $\mathbb{Z}$ and its arithmetic operation.

Remark 5.6 It is interesting to sort out the last example from the justification for the arithmetic of negative numbers that we learn in high school. Here is an example of that reasoning.

It is postulated that negative numbers exist (they certainly are not constructed). Then it is noted that

$$18 + (8 - 14) = (18 - 14) + 8 = 4 + 8 = 12 = 18 - 6 = 18 + (-6)$$

Identifying the far left and far right sides of the equation, we are forced to the conclusion that $8 - 14 = -6$.

This reasoning is perfectly correct. But it presupposes the existence of a number system that (*a*) contains negative integers and (*b*) obeys all the familiar laws of arithmetic.

The advantage of the presentation in this section of the present book is that we actually *construct* such a number system. We do not presuppose it. The additive properties of negative numbers follow automatically from our construction. They are not derived by algebraic tricks from some numbers that we do not actually know exist.

Finally we turn to multiplication. If $A = [(a, b)]$ and $C = [(c, d)]$ are integers then we define their product by the formula

$$A \cdot C = [(a \cdot d + b \cdot c, a \cdot c + b \cdot d)]$$

This definition may be a surprise. Why did we not define $A \cdot C$ to be $[(a \cdot c, b \cdot d)]$? There are several reasons: first of all, the latter definition would give the wrong answer; moreover, it is not unambiguous (different representatives of A and C would give a different answer). If you recall that we think of $[(a, b)]$ as representing $b - a$ and $[(c, d)]$ as representing $d - c$ then the product should be the equivalence class that represents $(b - a) \cdot (d - c)$. That is the motivation behind our definition.

The unambiguity (or well definedness) of the given definition of multiplication of integers is treated in the exercises. We proceed now to an example.

EXAMPLE 5.3
We compute the product of -3 and -6. Now

$$(-3) \cdot (-6) = [(5, 2)] \cdot [(9, 3)] = [(5 \cdot 3 + 2 \cdot 9, 5 \cdot 9 + 2 \cdot 3)]$$
$$= [(33, 51)] = 18$$

which is the expected answer.

As a second example, we multiply -5 and 12. We have

$$-5 \cdot 12 = [(7, 2)] \cdot [(1, 13)] = [(7 \cdot 13 + 2 \cdot 1, 7 \cdot 1 + 2 \cdot 13)]$$
$$= [(93, 33)] = -60$$

Finally, we show that 0 times any integer A equals zero. Let $A = [(a, b)]$. Then

$$0 \cdot A = [(1, 1)] \cdot [(a, b)] = [(1 \cdot b + 1 \cdot a, 1 \cdot a + 1 \cdot b)]$$
$$= [(a + b, a + b)] = 0 \qquad \square$$

Remark 5.7 Notice that one of the pleasant by-products of our construction of the integers is that we no longer have to give artificial explanations for why the product of two negative numbers is a positive number or why the product of a negative number and a positive number is negative. These properties instead follow *automatically* from our construction.

Remark 5.8 It is interesting to sort out the last example from the justification for the arithmetic of negative numbers that we learn in high school. Here is an example of that reasoning.

It is postulated that negative numbers exist (they certainly are not constructed). Then it is noted that

$$3 \cdot 8 = (6 - 3) \cdot 8 = 6 \cdot 8 - 3 \cdot 8$$

hence

$$24 = 48 - 3 \cdot 8$$

or, using reasoning as in our last remark but one,

$$-24 = -3 \cdot 8$$

Similarly, one can show that

$$-48 = -6 \cdot 8$$

Taking these two facts for granted, we then compute that

$$(8 - 3) \cdot (8 - 6) = 8 \cdot 8 + 8 \cdot (-6) + (-3) \cdot 8 + (-3) \cdot (-6)$$

As a result,

$$10 = 64 - 48 - 24 + (-3) \cdot (-6)$$

or

$$10 + 72 - 64 = (-3) \cdot (-6)$$

hence

$$18 = (-3) \cdot (-6)$$

Again, this reasoning is perfectly correct. But it presupposes the existence of a number system that (*a*) contains negative integers and (*b*) obeys all the familiar laws of arithmetic.

The advantage of the presentation in this section of the present book is that we actually *construct* such a number system. We do not presuppose it. The multiplicative properties of negative numbers follow automatically from our construction. They are not derived by algebraic tricks from some numbers that we do not actually know exist.

Notice that the integers $\mathbb{Z}$, as we have constructed them, contain the element $0 \equiv [(1, 1)]$. This element is the *additive identity* in the sense that $x + 0 = 0 + x = x$ for any integer x. Also, if $y = [(a, b)]$ is any integer then it has an *additive inverse* $-y = [(b, a)]$. This means that $y + (-y) = 0$. As a result of these two facts, the integers $\mathbb{Z}$ form a *group*. We shall say more about groups in Sec. 9.4.

Of course we will not discuss division for integers; in general division of one integer by another makes no sense *in the universe of the integers*. More will be said about this fact in the Exercises.

In the rest of this book we will follow the standard mathematical custom of denoting the set of all integers by the symbol $\mathbb{Z}$. We will write the integers not as equivalence classes, but in the usual way as the sequence of digits $\ldots -3, -2, -1, 0, 1, 2, 3, \ldots$. The equivalence classes are a device that we used to *construct* the integers. Now that we have them, we may as well write them in the simple, familiar fashion and manipulate them as usual.

In an exhaustive treatment of the construction of $\mathbb{Z}$, we would prove that addition and multiplication are commutative and associative, prove the distributive law, and so forth. But the purpose of this section is to demonstrate modes of logical thought rather than to be exhaustive. We shall say more about some of the elementary properties of the integers in the Exercises.

5.4 The Rational Numbers

In this section we use the integers, together with a construction using equivalence classes, to build the rational numbers. Let A be the set $\mathbb{Z} \times (\mathbb{Z} \setminus \{0\})$. In other words, A is the set of ordered pairs (a, b) of integers subject to the condition that $b \neq 0$. (*Think of this ordered pair as ultimately "representing" the fraction a/b.*) We definitely want it to be the case that certain ordered pairs represent the same number. For instance,

$$\frac{1}{2} \text{ should be the same number as } \frac{3}{6}$$

This motivates our equivalence relation. Declare (a, b) to be related to (a^*, b^*) if $a \cdot b^* = a^* \cdot b$. (*Here we are thinking that the fraction a/b should equal the fraction a^*/b^* precisely when $a \cdot b^* = a^* \cdot b$.*)

Is this an equivalence relation? Obviously the pair (a, b) is related to itself, since $a \cdot b = a \cdot b$. Also the relation is symmetric: if (a, b) and (a^*, b^*) are pairs and $a \cdot b^* = a^* \cdot b$ then $a^* \cdot b = a \cdot b^*$. Finally, if (a, b) is related to (a^*, b^*) and (a^*, b^*) is related to (a^{**}, b^{**}) then we have both

$$a \cdot b^* = a^* \cdot b \quad \text{and} \quad a^* \cdot b^{**} = a^{**} \cdot b^* \tag{5.1}$$

Multiplying the left sides of these two equations together and the right sides together gives

$$(a \cdot b^*) \cdot (a^* \cdot b^{**}) = (a^* \cdot b) \cdot (a^{**} \cdot b^*)$$

If $a^* = 0$ then it follows immediately from Eq. 5.1 that both a and a^{**} must be zero. So the three pairs (a, b), (a^*, b^*), and (a^{**}, b^{**}) are equivalent and there is nothing to prove. So we may assume that $a^* \neq 0$. We know a priori that $b^* \neq 0$; therefore we may cancel common terms in the last equation to obtain

$$a \cdot b^{**} = b \cdot a^{**}$$

Thus (a, b) is related to (a^{**}, b^{**}), and our relation is transitive. (*Exercise*: Explain why it is correct to "cancel common terms" in the last step.)

The resulting collection of equivalence classes will be called the set of *rational numbers*, and we shall denote this set with the symbol $\mathbb{Q}$.

EXAMPLE 5.4
The equivalence class [(4, 12)] contains all of the pairs (4, 12), (1, 3), (−2, −6). (Of course it contains infinitely many other pairs as well.) This equivalence class represents the fraction 4/12 which we sometimes also write as 1/3 or (−2)/(−6). □

If $[(a, b)]$ and $[(c, d)]$ are rational numbers then we define their *product* to be the rational number

$$[(a \cdot c, b \cdot d)]$$

This is well defined (unambiguous), for the following reason. Suppose that (a, b) is related to $(\tilde{a}, \tilde{b})$ and (c, d) is related to $(\tilde{c}, \tilde{d})$. We would like to know that $[(a, b)] \cdot [(c, d)] = [(a \cdot c, b \cdot b)]$ is the same equivalence class as $[(\tilde{a}, \tilde{b})] \cdot [(\tilde{c}, \tilde{d})] = [(\tilde{a} \cdot \tilde{c}, \tilde{b} \cdot \tilde{d})]$. In other words we need to know that

$$(a \cdot c) \cdot (\tilde{b} \cdot \tilde{d}) = (\tilde{a} \cdot \tilde{c}) \cdot (b \cdot d). \tag{5.2}$$

But our hypothesis is that

$$a \cdot \tilde{b} = \tilde{a} \cdot b \quad \text{and} \quad c \cdot \tilde{d} = \tilde{c} \cdot d$$

Multiplying together the left sides and the right sides we obtain

$$(a \cdot \tilde{b}) \cdot (c \cdot \tilde{d}) = (\tilde{a} \cdot b) \cdot (\tilde{c} \cdot d)$$

Rearranging, we have

$$(a \cdot c) \cdot (\tilde{b} \cdot \tilde{d}) = (\tilde{a} \cdot \tilde{c}) \cdot (b \cdot d)$$

But this is just Eq. 5.2. So multiplication is unambiguous.

EXAMPLE 5.5
The product of the two rational numbers [(3, 8)] and [(−2, 5)] is

$$[(3 \cdot (-2), 8 \cdot 5)] = [(-6, 40)] = [(-3, 20)]$$

This is what we expect: the product of 3/8 and −2/5 is −3/20. □

If $q = [(a, b)]$ and $r = [(c, d)]$ are rational numbers and if r is not zero (that is, $[(c, d)]$ is not the equivalence class zero—in other words, $c \neq 0$) then we define the quotient q/r to be the equivalence class

$$[(ad, bc)]$$

We leave it to you to check that this operation is well defined.

EXAMPLE 5.6
The quotient of the rational number $[(4, 7)]$ by the rational number $[(3, -2)]$ is, by definition, the rational number

$$[(4 \cdot (-2), 7 \cdot 3)] = [(-8, 21)]$$

This is what we expect: the quotient of 4/7 by −3/2 is −8/21.   □

How should we add two rational numbers? We could try declaring $[(a, b)] + [(c, d)]$ to be $[(a + c, b + d)]$, but this will not work (think about the way that we usually add fractions). Instead we define

$$[(a, b)] + [(c, d)] = [(a \cdot d + b \cdot c, b \cdot d)]$$

That this definition is well defined (unambiguous) is left for the exercises. We turn instead to an example.

EXAMPLE 5.7
The sum of the rational numbers $[(3, -14)]$ and $[(9, 4)]$ is given by

$$[(3 \cdot 4 + (-14) \cdot 9, (-14) \cdot 4)] = [(-114, -56)] = [(57, 28)]$$

This coincides with the usual way that we add fractions:

$$-\frac{3}{14} + \frac{9}{4} = \frac{57}{28}$$ □

Notice that the equivalence class $[(0, 1)]$ is the rational number that we usually denote by 0. It is the additive identity, for if $[(a, b)]$ is another rational number then

$$[(0, 1)] + [(a, b)] = [(0 \cdot b + 1 \cdot a, 1 \cdot b)] = [(a, b)]$$

A similar argument shows that $[(0, 1)]$ times any rational number gives $[(0, 1)]$ or 0. By the same token the rational number $[(1, 1)]$ is the multiplicative identity. We leave the details for you.

Of course the concept of subtraction is really just a special case of addition [that is, $a - b$ is the same as $a + (-b)$]. So we shall say nothing further about subtraction.

In practice we will write rational numbers in the traditional fashion:

$$\frac{2}{5}, \frac{-19}{3}, \frac{22}{2}, \frac{24}{4}, \dots$$

In mathematics it is generally not wise to write rational numbers in mixed form, such as $2\frac{3}{5}$, because the juxtaposition of two numbers could easily be mistaken for multiplication. Instead we would write this quantity as the improper fraction 13/5.

Definition 5.1 A set S is called a *field* if it is equipped with a binary operation (usually called addition and denoted "+") and a second binary operation (usually called multiplication and denoted "·") such that the following axioms are satisfied:

A1. S is closed under addition: if $x, y \in S$ then $x + y \in S$.

A2. Addition is commutative: if $x, y \in S$ then $x + y = y + x$.

A3. Addition is associative: if $x, y, z \in S$ then

$$x + (y + z) = (x + y) + z.$$

A4. There exists an element, called 0, in S which is an additive identity: if $x \in S$ then $0 + x = x$.

A5. Each element of S has an additive inverse: if $x \in S$ then there is an element $-x \in S$ such that $x + (-x) = 0$.

M1. S is closed under multiplication: if $x, y \in S$ then $x \cdot y \in S$.

M2. Multiplication is commutative: if $x, y \in S$ then $x \cdot y = y \cdot x$.

M3. Multiplication is associative: if $x, y, z \in S$ then $x \cdot (y \cdot z) = (x \cdot y) \cdot z$.

M4. There exists an element, called 1, which is a multiplicative identity: if $x \in S$ then $1 \cdot x = x$.

M5. Each nonzero element of S has a multiplicative inverse: if $0 \neq x \in S$ then there is an element $x^{-1} \in S$ such that $x \cdot (x^{-1}) = 1$. The element x^{-1} is sometimes denoted by $1/x$.

D1. Multiplication distributes over addition: if $x, y, z \in S$ then $x \cdot (y + z) = x \cdot y + x \cdot z$.

Eleven axioms is a lot to digest all at once, but in fact these are all familiar properties of addition and multiplication of rational numbers that we use every day: the set $\mathbb{Q}$, with the usual notions of addition and multiplication (and with the usual additive identity 0 and multiplicative identity 1), forms a field. The integers, by contrast, do not: nonzero elements of $\mathbb{Z}$ (except 1 and -1) do not have multiplicative inverses *in the integers*.

Let us now consider some consequences of the field axioms.

Theorem 5.3 *Any field has the following properties:*

(1) *If* $z + x = z + y$ *then* $x = y$.

(2) *If* $x + z = 0$ *then* $z = -x$ *(the additive inverse is unique).*

(3) $-(-y) = y$.

(4) *If* $y \neq 0$ *and* $y \cdot x = y \cdot z$ *then* $x = z$.

(5) *If* $y \neq 0$ *and* $y \cdot z = 1$ *then* $z = y^{-1}$ *(the multiplicative inverse is unique).*

(6) $(x^{-1})^{-1} = x$.

(7) $0 \cdot x = 0$.

(8) *If* $x \cdot y = 0$ *then either* $x = 0$ *or* $y = 0$.

(9) $(-x) \cdot y = -(x \cdot y) = x \cdot (-y)$.

(10) $(-x) \cdot (-y) = x \cdot y$.

Proof: These are all familiar properties of the rationals, but now we are considering them for an arbitrary field. We prove just a few to illustrate the logic. The proofs of the others are assigned as exercises.

To prove **(1)** we write

$$z + x = z + y \Rightarrow (-z) + (z + x) = (-z) + (z + y)$$

and now Axiom **A3** yields that this implies

$$[(-z) + z] + x = [(-z) + z] + y$$

Next, Axiom **A5** yields that

$$0 + x = 0 + y$$

and hence, by Axiom **A4**,

$$x = y$$

To prove **(7)**, we observe that

$$0 \cdot x = (0 + 0) \cdot x$$

which by Axiom **M2** equals

$$x \cdot (0 + 0)$$

By Axiom **D1** the last expression equals

$$x \cdot 0 + x \cdot 0$$

which by Axiom **M2** equals $0 \cdot x + 0 \cdot x$. Thus we have derived the equation

$$0 \cdot x = 0 \cdot x + 0 \cdot x$$

Axioms **A4** and **A2** let us rewrite the left side as

$$0 \cdot x + 0 = 0 \cdot x + 0 \cdot x$$

Finally, part **(1)** of the present theorem (which we have already proved) yields that

$$0 = 0 \cdot x$$

which is the desired result.

To prove **(8)**, we suppose that $x \neq 0$. In this case x has a multiplicative inverse x^{-1} and we multiply both sides of our equation by this element:

$$x^{-1} \cdot (x \cdot y) = x^{-1} \cdot 0$$

By Axiom **M3**, the left side can be rewritten and we have

$$(x \cdot x^{-1}) \cdot y = x^{-1} \cdot 0$$

Next, we rewrite the right side using Axiom **M2**:

$$(x \cdot x^{-1}) \cdot y = 0 \cdot x^{-1}$$

Now Axiom **M5** allows us to simplify the left side:

$$1 \cdot y = 0 \cdot x^{-1}$$

We further simplify the left side using Axiom **M4** and the right side using **(7)** of the present theorem (which we just proved) to obtain:

$$y = 0$$

Thus we see that if $x \neq 0$ then $y = 0$. But this is logically equivalent with $x = 0$ or $y = 0$, as we wished to prove. (If you have forgotten why these statements are logically equivalent, write out a truth table.) □

EXAMPLE 5.8
The integers $\mathbb{Z}$ form a strictly, simply ordered set when equipped with the usual ordering. We can make this ordering precise by saying that $x < y$ if $y - x$ is a positive integer. For instance,

$$6 < 8 \quad \text{because} \quad 8 - 6 = 2 > 0$$

Likewise,

$$-5 < -1 \quad \text{because} \quad -1 - (-5) = 4 > 0$$

Observe that the same ordering works on the rational numbers. □

If A is a strictly ordered set and a, b are elements then we often write $a \leq b$ to mean that *either $a = b$ or $a < b$.*

When a field has an ordering which is compatible with the field operations then a richer structure results.

Definition 5.2 A field F is called an *ordered field* if F has a strict, simple ordering $<$ that satisfies the following addition properties:

1. If $x, y, z \in F$ and if $y < z$ then $x + y < x + z$.
2. If $x, y \in F$, $x > 0$, and $y > 0$ then $x \cdot y > 0$.

Again, these are familiar properties of the rational numbers: $\mathbb{Q}$ forms an ordered field. Some further properties of ordered fields may be proved from the axioms.

Theorem 5.4 *Any ordered field has the following properties:*

(1) *If $x > 0$ and $z < y$ then $x \cdot z < x \cdot y$.*
(2) *If $x < 0$ and $z < y$ then $x \cdot z > x \cdot y$.*
(3) *If $x > 0$ then $-x < 0$. If $x < 0$ then $-x > 0$.*
(4) *If $0 < y < x$ then $0 < 1/x < 1/y$.*
(5) *If $x \neq 0$ then $x^2 > 0$.*
(6) *If $0 < x < y$ then $x^2 < y^2$.*

Proof: Again we prove just a few of these statements and leave the rest as exercises.

To prove **(1)**, observe that the property **(1)** of ordered fields together with our hypothesis implies that

$$(-z) + z < (-z) + y$$

Thus, using Axiom **A2**, we see that $y - z > 0$. Since $x > 0$, property **(2)** of ordered fields gives

$$x \cdot (y - z) > 0$$

Finally,

$$x \cdot y = x \cdot [(y - z) + z] = x \cdot (y - z) + x \cdot z > 0 + x \cdot z$$

[by property (**1**) of ordered fields again]. In conclusion,

$$x \cdot y > x \cdot z$$

To prove (**3**), begin with the equation

$$0 = -x + x$$

Since $x > 0$, the right side is greater than $-x$. Thus $0 > -x$ as claimed. The proof of the other statement of (**3**) is similar.

To prove (**5**), we consider two cases. If $x > 0$ then $x^2 \equiv x \cdot x$ is positive by property (**2**) of ordered fields. If $x < 0$ then $-x > 0$ [by part (**3**) of the present theorem, which we just proved] hence $(-x) \cdot (-x) > 0$. But part (**10**) of the last theorem guarantees that $(-x) \cdot (-x) = x \cdot x$ hence $x \cdot x > 0$. □

5.5 The Real Number System

Now that we are accustomed to the notion of equivalence classes, the construction of the integers and of the rational numbers seems fairly natural. In fact equivalence classes provide a precise language for declaring certain objects to be equal (or for identifying certain objects). We can now use the integers and the rationals as we always have done, with the added confidence that they are not simply a useful notation but that they have been *constructed*.

We turn next to the real numbers. We saw in Sec. 5.4 that the rational number system is not closed under the operation of taking square roots, for example. We know from calculus that for many other purposes the rational numbers are inadequate. It is important to work in a number system which is closed with respect to all the operations we shall perform. While the rationals are closed under the usual arithmetic operations, they are not closed under the operation of taking *limits*. For instance, the sequence of rational numbers 3, 3.1, 3.14, 3.141, . . . consists of terms that seem to be getting closer and closer together, *seem* to tend to some limit, and yet there is no rational number which will serve as a limit (of course it turns out that the limit is π—an "irrational" number).

We will now deal with the real number system, a system which contains all limits of sequences of rational numbers (as well as all limits of sequences of real numbers!). In fact our plan will be as follows: in this section we shall discuss all the requisite properties of the reals. The actual construction of the reals is rather complicated, and we shall put that in an appendix (Sec. 5.5.1) at the end of the section.

Definition 5.3 Let A be an ordered set and X a subset of A. The set X is called *bounded above* if there is an element $b \in A$ such that $x \leq b$ for all $x \in X$. We call the element b an *upper bound* for the set X.

EXAMPLE 5.9
Let $A = \mathbb{Q}$ with the usual ordering. The set $X = \{x \in \mathbb{Q} : 2 < x < 4\}$ is bounded above. For example, the number 15 is an upper bound for X. So are the numbers 12 and 4. It is interesting to observe that no element of this particular X can actually be an upper bound for X. The number 4 is a good candidate, but 4 is not an element of X. In fact if $b \in X$ then $(b + 4)/2 \in X$ and $b < (b + 4)/2$, so b could not be an upper bound for X. □

It turns out that the most convenient way to formulate the notion that the real numbers have "no gaps" (that is, that all sequences which seem to be converging actually have something to converge to) is in terms of upper bounds.

Definition 5.4 Let A be an ordered set and X a subset of A. An element $b \in A$ is called a *least upper bound* (or *supremum*) for X if b is an upper bound for X and there is no upper bound b^* for X which is less than b.

By its very definition, if a least upper bound exists then it is unique.

EXAMPLE 5.10
In the last example, we considered the set X of rational numbers strictly between 2 and 4. We observed there that 4 is the least upper bound for X. Note that this least upper bound is not an element of the set X.

The set $Y = \{y \in \mathbb{Z} : -9 \le y \le 7\}$ has least upper bound 7. In this case, the least upper bound *is* an element of the set Y. □

Notice that we may define a lower bound for a subset of an ordered set in a fashion similar to that for an upper bound: $l \in A$ is a lower bound for $X \subset A$ if $x \ge l$ for all $x \in X$. A *greatest lower bound* (or *infimum*) for X is then defined to be a lower bound ℓ such that there is no lower bound ℓ^* with $\ell^* > \ell$.

EXAMPLE 5.11
The set X in the last two examples has lower bounds $-20, 0, 1, 2$, for instance. The greatest lower bound is 2, which is *not* an element of the set.

The set Y in the last example has lower bounds $-53, -22, -10, -9$, to name just a few. The number -9 is the greatest lower bound. It *is* an element of Y. □

EXAMPLE 5.12
Let $S = \mathbb{Z} \subset \mathbb{R}$. Then S does not have an upper bound. □

The purpose that the real numbers will serve for us is as follows: they will contain the rationals, they will still be an ordered field, and *every subset which has an upper bound will have a least upper bound*. We formulate this property as a theorem.

Theorem 5.5 *There exists an ordered field* $\mathbb{R}$ *which (a) contains* $\mathbb{Q}$ *and (b) has the property that any nonempty subset of* $\mathbb{R}$ *which has an upper bound has a least upper bound.*

The last property described in this theorem is called the *least upper bound property* of the real numbers. As mentioned previously, this theorem will be proved in Sec. 5.5.1. Now we begin to realize why it is so important to *construct* the number systems that we will use. We are endowing $\mathbb{R}$ with a great many properties. Why do we have any right to suppose that there exists a number system with all these properties? We must produce one!

Let us begin to explore the richness of the real numbers. The next theorem states a property which is certainly not shared by the rationals (see Sec. 5.4). It is fundamental in its importance.

Theorem 5.6 *Let* x *be a real number such that* $x > 0$. *Then there is a positive real number* y *such that* $y^2 = y \cdot y = x$.

Proof: We will use throughout this proof the fact (see part (**6**) of Theorem 5.4) that if $0 < a < b$ then $a^2 < b^2$.

Let

$$S = \{s \in \mathbb{R} : s > 0 \text{ and } s^2 < x\}$$

Then S is not empty since $x/2 \in S$ if $x < 2$ and $1 \in S$ otherwise. Also S is bounded above since $x + 1$ is an upper bound for S. By Theorem 5.5, the set S has a least upper bound. Call it y. Obviously $0 < \min\{x/2, 1\} \le y$ hence y is positive. We claim that $y^2 = x$. To see this, we eliminate the other two possibilities.

If $y^2 < x$ then set $\varepsilon = (x - y^2)/[4(x + 1)]$. Then $\varepsilon > 0$ and

$$
\begin{aligned}
(y + \varepsilon)^2 &= y^2 + 2 \cdot y \cdot \varepsilon + \varepsilon^2 \\
&= y^2 + 2 \cdot y \cdot \frac{x - y^2}{4(x + 1)} + \frac{x - y^2}{4(x + 1)} \cdot \frac{x - y^2}{4(x + 1)} \\
&< y^2 + 2 \cdot \frac{y}{x + 1} \cdot \frac{x - y^2}{4} + \frac{x - y^2}{4} \cdot \frac{x}{4x} \\
&< y^2 + \frac{x - y^2}{2} + \frac{x - y^2}{16} \\
&< y^2 + (x - y^2) \\
&= x
\end{aligned}
$$

Thus $y + \varepsilon \in S$, and y cannot be an upper bound for S. This contradiction tells us that $y^2 \not< x$.

Similarly, if it were the case that $y^2 > x$ then we set $\varepsilon = (y^2 - x)/[4(x + 1)]$. A calculation like the one we just did then shows that $(y - \varepsilon)^2 \geq x$. Hence $y - \varepsilon$ is also an upper bound for S, and y is therefore not the *least* upper bound. This contradiction shows that $y^2 \not> x$.

The only remaining possibility is that $y^2 = x$.   □

A similar proof shows that if n is a positive integer and x a positive real number then there is a positive real number y such that $y^n = x$.

We next use the least upper bound property of the real numbers to establish two important qualitative properties of the real numbers:

Theorem 5.7 *The set* $\mathbb{R}$ *of real numbers satisfies the archimedean property:*

"Let a and b be positive real numbers. Then there is a natural number n such that
*   na > b."*

Theorem 5.8 *The set* $\mathbb{Q}$ *of rational numbers satisfies the following density property:*

"Let c < d be real numbers. Then there is a rational number q with c < q < d."

Proof of Theorem 5.7:   Suppose the archimedean property to be false. Then $S = \{na : n \in \mathbb{N}\}$ has b as an upper bound. Therefore S has a finite supremum β. Since $a > 0$, $\beta - a < \beta$. So $\beta - a$ is not an upper bound for S, and there must be a natural number n^* such that $n^* \cdot a > \beta - a$. But then $(n^* + 1)a > \beta$, and β cannot be the supremum for S. This contradiction proves the theorem.   □

Proof of Theorem 5.8:   Let $\lambda = d - c > 0$. By the archimedean property, choose a positive integer N such that $N \cdot \lambda > 1$. Again the archimedean property gives a natural number P such that $P > N \cdot c$ and another Q such that $Q > |-N \cdot c|$. Then $Q > -N \cdot c$ and we see that Nc falls between the integers $-Q$ and P; therefore there must be an integer M between $-Q$ and P (inclusive) such that

$$M - 1 \leq Nc < M$$

Thus $c < M/N$. Also

$$M \leq Nc + 1$$

hence

$$\frac{M}{N} \leq c + \frac{1}{N} < c + \lambda = d$$

So M/N is a rational number lying strictly between c and d.   □

One of the most profound and useful properties of the real numbers, and one that is equivalent to the least upper bound property, is the *intermediate value property*:

Theorem 5.9 *Let f be a continuous, real-valued function with domain the interval $[a, b]$. If $f(a) = \alpha$, $f(b) = \beta$, and if $\alpha < \gamma < \beta$ then there is a value $t_0 \in (a, b)$ such that $f(t_0) = \gamma$.*

Proof: Let

$$S = \{x \in [a, b] : f(x) < \gamma\}$$

Then $S \neq \emptyset$ since $a \in S$. Moreover S is bounded above by b. So $t_0 = \sup S$ exists as a finite real number. We claim that $f(t_0) = \gamma$.

Clearly $f(t_0) \leq \gamma$ since t_0 is the limit of numbers at which f takes values less than γ (we use the continuity of f here). Suppose, seeking a contradiction, that $f(t_0) < \gamma$. Let $\epsilon = \gamma - f(t_0)$. By the continuity of f, we may select $\delta > 0$ such that $|t - t_0| < \delta$ implies that $|f(t) - f(t_0)| < \epsilon/2$. But then, for $t \in (t_0 - \delta, t_0 + \delta)$, $f(t) < f(t_0) + \epsilon/2 < \gamma$. It follows that $(t_0 - \delta, t_0 + \delta) \subset S$, so t_0 cannot be the supremum of S. That is a contradiction. Therefore $f(t_0) = \gamma$. □

As an application, we prove the following special case of a theorem of Brouwer:

Theorem 5.10 *Let $f : [0, 1] \to [0, 1]$ be a continuous function. Then f has a fixed point, in the sense that there is a point $c \in [0, 1]$ such that $f(c) = c$.*

Proof: Seeking a contradiction, we suppose not. Then, in particular, $f(0) > 0$ and $f(1) < 1$. Now set $g(x) = x - f(x)$. We see that $g(0) = 0 - f(0) < 0$ and $g(1) = 1 - f(1) > 0$. By the intermediate value property, there must therefore be a point c between 0 and 1 such that $g(c) = 0$. But this says that $f(c) = c$, as required. □

We conclude by recalling the "absolute value" notation.

Definition 5.5 Let x be a real number. We define

$$|x| = \begin{cases} x & \text{if } x > 0 \\ 0 & \text{if } x = 0 \\ -x & \text{if } x < 0 \end{cases}$$

The absolute value of a real number x measures the distance of x to 0. It is left as an exercise for you to verify the important *triangle inequality*:

$$|x + y| \leq |x| + |y|$$

5.5.1 CONSTRUCTION OF THE REAL NUMBERS

There are several techniques for constructing the real number system $\mathbb{R}$ from the rational number system $\mathbb{Q}$. We use the method of Dedekind (Julius W. R. Dedekind, 1831–1916) cuts because it uses a minimum of new ideas and is fairly brief.

Keep in mind that, throughout this appendix, our universe is the system of rational numbers $\mathbb{Q}$. We are *constructing* the new number system $\mathbb{R}$.

Definition 5.6 A *cut* is a subset C of $\mathbb{Q}$ with the following properties:

1. $C \neq \emptyset$
2. If $s \in C$ and $t < s$ then $t \in C$
3. If $s \in C$ then there is a $u \in C$ such that $u > s$
4. There is a rational number x such that $c < x$ for all $c \in C$

You should think of a cut C as the set of all rational numbers to the left of some point in the real line (that is, it is an open half-line of rational numbers). For example, the set $\{x \in \mathbb{Q} : x^2 < 2\} \cup \{x \in \mathbb{Q} : x < 0\}$ is a cut. Roughly speaking, it is the set of rational numbers to the left of $\sqrt{2}$. (Take care to note that $\sqrt{2}$ does not exist as a rational number; so we are using a circuitous method to specify this set.) Since we have not constructed the real line yet, we cannot define this cut in that simple way; we have to make the construction more indirect. But if you consider the four properties of a cut, they describe a set that looks like a "rational left half-line."

Notice that if C is a cut and $s \notin C$ then any rational $t > s$ is also not in C. Also, if $r \in C$ and $s \notin C$ then it must be that $r < s$.

Definition 5.7 If C and D are cuts then we say that $C < D$ provided that C is a subset of D but $C \neq D$.

Check for yourself that "$<$" is a strict, simple ordering on the set of all cuts. We note that $C = D$ if and only if $C \subset D$ and $D \subset C$.

Now we introduce operations of addition and multiplication which will turn the set of all cuts into a field.

Definition 5.8 If C and D are cuts then we define

$$C + D = \{c + d : c \in C, d \in D\}$$

We define the cut $\hat{0}$ to be the set of all negative rationals.

The cut $\hat{0}$ will play the role of the additive identity. We are now required to check that field Axioms **A1–A5** hold.

Discrete Mathematics Demystified

For **A1**, we need to see that $C + D$ is a cut. Obviously $C + D$ is not empty. If s is an element of $C + D$ and t is a rational number less than s, write $s = c + d$, where $c \in C$ and $d \in D$. Then $t - c < s - c = d \in D$ so $t - c \in D$; and $c \in C$. Hence $t = c + (t - c) \in C + D$. A similar argument shows that there is an $r > s$ such that $r \in C + D$. Finally, if x is a rational upper bound for C and y is a rational upper bound for D, then $x + y$ is a rational upper bound for $C + D$. We conclude that $C + D$ is a cut.

Since addition of rational numbers is commutative, it follows immediately that addition of cuts is commutative. Associativity follows in a similar fashion. That takes care of **A2** and **A3**.

Now we show that if C is a cut then $C + \hat{0} = C$. For if $c \in C$ and $z \in \hat{0}$ then $c + z < c + 0 = c$ hence $C + \hat{0} \subset C$. Also, if $c^* \in C$ then choose a $d^* \in C$ such that $c^* < d^*$. Then $c^* - d^* < 0$ so $c^* - d^* \in \hat{0}$. And $c^* = d^* + (c^* - d^*)$. Hence $C \subset C + \hat{0}$. We conclude that $C + \hat{0} = C$. This is **A4**.

Finally, for Axiom **A5**, we let C be a cut and set $-C$ to be equal to $\{d \in \mathbb{Q} : \exists d^* > d$ such that $c + d^* < 0$ for all $c \in C\}$. If x is a rational upper bound for C then $-x \in -C$ so $-C$ is not empty. It is also routine to check that $-C$ is a cut. By its very definition, $C + (-C) \subset \hat{0}$.

Further, if $z \in \hat{0}$ then there is a $z^* \in \hat{0}$ such that $z < z^*$. Choose an element $c \in C$ such that $c + (z^* - z) \notin C$ (why is this possible?). Let $c^* \in C$ be such that $c < c^*$. Set $c^{**} = z - c^*$. Then $d^* = z - c > c^{**}$. We claim that $\tilde{c} + d^* < 0$ for all $\tilde{c} \in C$. Suppose for the moment that this claim has been proved. Then this shows that $c^{**} \in -C$. Then $z = c^* + c^{**} \in C + (-C)$ so that $\hat{0} \subset C + (-C)$. We then conclude that $C + (-C) = \hat{0}$, and Axiom **A5** is established.

It remains to prove the claim. So let d^* be defined as above and select $\tilde{c} \in C$. Then

$$d^* + \tilde{c} = z + (-c + \tilde{c}) < z + (z^* - z) = z^* < 0$$

Here we have used the choice of c. This establishes the claim and completes the proof of **A5**.

Having verified the axioms for addition, we turn now to multiplication.

Definition 5.9 If C and D are cuts then we define the product $C \cdot D$ as follows:

- If $C, D > \hat{0}$ then $C \cdot D = \{q \in \mathbb{Q} : q < c \cdot d$ for some $c \in C, d \in D$ with $c > 0, d > 0\}$
- If $C > \hat{0}, D < \hat{0}$ then $C \cdot D = -[C \cdot (-D)]$
- If $C < \hat{0}, D > \hat{0}$ then $C \cdot D = -[(-C) \cdot D]$
- If $C, D < \hat{0}$ then $C \cdot D = (-C) \cdot (-D)$
- If either $C = \hat{0}$ or $D = \hat{0}$ then $C \cdot D = \hat{0}$

Notice that, for convenience, we have defined multiplication of negative numbers just as we did in high school. The reason is that the definition that we use for the product of two positive numbers cannot work when one of the two factors is negative (check this as an exercise).

We have said what the additive identity is in this realization of the real numbers. Of course the multiplicative identity is the cut corresponding to 1, or

$$\widehat{1} \equiv \{t \in \mathbb{Q} : t < 1\}$$

We leave it to the reader to verify that if C is any cut then $\widehat{1} \cdot C = C \cdot \widehat{1} = C$.

It is now routine to verify that the set of all cuts, with this definition of multiplication, satisfies field Axioms **M1–M5**. The proofs follow those for **A1–A5** rather closely.

For the distributive property, one first checks the case when all the cuts are positive, reducing it to the distributive property for the rationals. Then one handles negative cuts on a case-by-case basis.

The two properties of an ordered field are also easily checked for the set of all cuts.

We now know that the collection of all cuts forms an ordered field. Denote this field by the symbol $\mathbb{R}$ and call it the real number system. We next verify the crucial property of $\mathbb{R}$ that sets it apart from $\mathbb{Q}$.

Theorem 5.11 *The ordered field $\mathbb{R}$ satisfies the least upper bound property.*

Proof: Let S be a subset of $\mathbb{R}$ which is bounded above. That is, there is a cut α such that $s < \alpha$ for all $s \in S$. Define

$$S^* = \bigcup_{C \in S} C$$

Then S^* is clearly nonempty, and it is therefore a cut since it is a union of cuts. It is also clearly an upper bound for S since it contains each element of S. It remains to check that S^* is the least upper bound for S.

In fact if $T < S^*$ then $T \subset S^*$ and there is a rational number q in $S^* \setminus T$. But, by the definition of S^*, it must be that $q \in C$ for some $C \in S$. So $C > T$, and T cannot be an upper bound for S. Therefore S^* is the least upper bound for S, as desired. $\square$

We have shown that $\mathbb{R}$ is an ordered field which satisfies the least upper bound property. It remains to show that $\mathbb{R}$ contains (a copy of) $\mathbb{Q}$ in a natural way. In fact, if $q \in \mathbb{Q}$ we associate to it the element $\varphi(q) = C_q \equiv \{x \in \mathbb{Q} : x < q\}$. Then C_q is obviously a cut. It is also routine to check that

$$\varphi(q + q^*) = \varphi(q) + \varphi(q^*) \quad \text{and} \quad \varphi(q \cdot q^*) = \varphi(q) \cdot \varphi(q^*)$$

Therefore we see that ϕ is a ring homomorphism (see [LAN]) and hence represents $\mathbb{Q}$ as a "subfield" of $\mathbb{R}$.

5.6 The Nonstandard Real Number System

5.6.1 THE NEED FOR NONSTANDARD NUMBERS

Isaac Newton's calculus was premised on the existence of certain "infinitesimal numbers"—numbers that are positive, smaller than any standard real number, but not zero. Since limits were not understood in Newton's time, infinitesimals served in their stead. But in fact it was just these infinitesimals that called the theory of calculus into doubt. More than a century was expended developing the theory of limits in order to dispel those doubts.

Nonstandard analysis, due to Abraham Robinson (1918–1974), is a model for the real numbers (that is, it is a number system that satisfies the axioms for the real numbers that we enunciated in Sec. 5.5) that also contains infinitesimals. In a sense, then, Robinson's nonstandard reals are a perfectly rigorous theory that vindicates Newton's original ideas about infinitesimally small numbers.

5.6.2 FILTERS AND ULTRAFILTERS

One of the most standard constructions of the nonstandard real numbers involves putting an equivalence relation on the set of all sequences $\{a_j\}$ of real numbers. A natural algebraic construction for doing so is the *ultrafilter*. In fact ultrafilters are widely used in model theory (see the article by P. C. Eklof in [BAR]). So we will briefly say now what an ultrafilter is.

Let I be a nonempty set. A *filter* over I is a set $D \subseteq \mathcal{P}(I)$ such that

1. $\emptyset \notin D, I \in D$;
2. If $X, Y \in D$ then $X \cap Y \in D$;
3. If $X \in D$ and $X \subseteq Y \subseteq I$ then $Y \in D$.

In particular, a filter D over I has the *finite intersection property*: the intersection of any finite set of elements of D is nonempty.

A filter D over I is called an *ultrafilter* if, for every $X \subseteq I$, either $X \in D$ or $I \setminus X \in D$. It turns out that a filter over I is an ultrafilter if and only if it is a maximal filter over I (that is, there is no larger filter containing it). One can show, using Zorn's lemma, that if S is a collection of subsets of I which has the finite intersection property then S is contained in an ultrafilter over I.

5.6.3 A USEFUL MEASURE

We will follow the exposition that may be found at

> `http://members.tripod.com/PhilipApps/howto.html`

See also [LIN], [CUT]. At the end, we will point out the ultrafilter that is lurking in the background.

Let m be a finitely additive measure on the set $\mathbb{N}$ of natural numbers such that

1. For any subset $A \subseteq \mathbb{N}$, $m(A)$ is either 0 or 1.
2. It holds that $m(\mathbb{N}) = 1$ and $m(B) = 0$ for any finite set B.

That such a measure m exists is an easy exercise with the Axiom of Choice.[1] We leave the details to the interested reader.

5.6.4 AN EQUIVALENCE RELATION

Let

$$S = \left\{ \{a_n\}_{n=1}^{\infty} : a_n \in \mathbb{R} \text{ for all } n = 1, 2, \ldots \right\}$$

Define a relation $\sim$ on S by

$$\{a_n\} \sim \{b_n\} \qquad \text{if and only if} \qquad m\{n : a_n = b_n\} = 1$$

Then $\sim$ is clearly an equivalence relation. We let $\mathbb{R}^* = S/\sim$ be the nonstandard real number system.[2] In other words $\mathbb{R}^*$ is the collection of equivalence classes induced by this equivalence relation.

---

[1]The rather innocent-sounding Axiom of Choice says that, given any collection of sets, there is a function that assigns to each set one of its elements. This axiom was first formulated by Ernst Zermelo (1871–1953) in 1904. It turns out to harbor many mysteries, and has had a profound influence on the development of modern mathematics.

[2]In fact this is the point where we use an ultrafilter. The set $\mathcal{M} = \{A \subseteq \mathbb{N} : m(A) = 1\}$ is an ultrafilter. We are moding out by this ultrafilter.

We let $[\{a_n\}]$ denote the equivalence class containing the sequence $\{a_n\}$. Then we define some of the elementary operations on $\mathbb{R}^*$ by

$$[\{a_n\}] + [\{b_n\}] = [\{a_n + b_n\}]$$

$$[\{a_n\}] \cdot [\{b_n\}] = [\{a_n \cdot b_n\}]$$

$$[\{a_n\}] < [\{b_n\}] \quad \text{iff} \quad m(\{n : a_n < b_n\}) = 1$$

Further, we identify a *standard real number* b with the equivalence class $[\{b, b, b, \ldots\}]$.

5.6.5 AN EXTENSION OF THE REAL NUMBER SYSTEM

We have seen that $\mathbb{R}^*$ clearly contains $\mathbb{R}$ in a natural way. And it contains other elements too. We call $x \in \mathbb{R}^*$ an *infinitesimal* if and only if $a \neq 0$ and $-a < x < a$ for every positive real number a. For example, $[\{1, 2/3, 1/3, \ldots\}]$ is an infinitesimal. We call $y \in \mathbb{R}^*$ an *infinitary number* if $y > b$ for every real number b or $y < d$ for every real number d. As an instance, $[\{1, 2, 3, \ldots\}]$ is an infinitary number.

It would be inappropriate in a book of this type to delve very far into the theory of the nonstandard reals. But at least now the reader has an idea of what the nonstandard real numbers are, and of how a number system could contain both the standard reals and also infinitesimals and infinitaries.

5.7 The Complex Numbers

When we first learn about the complex numbers, the most troublesome point is the very beginning: "Let's pretend that the number -1 has a square root. Call it i." What gives us the right to "pretend" in this fashion? The answer is that we have no such right. If -1 has a square root, we should be able to construct a number system in which that is the case. That is what we shall do in this section.

Definition 5.10 The system of *complex numbers*, denoted by the symbol $\mathbb{C}$, consists of all ordered pairs (a, b) of real numbers (in other words, $\mathbb{C} = \mathbb{R} \times \mathbb{R}$). We add two complex numbers (a, b) and (a^*, b^*) by the formula

$$(a, b) + (a^*, b^*) = (a + a^*, b + b^*)$$

We multiply two complex numbers by the formula

$$(a, b) \cdot (a^*, b^*) = (a \cdot a^* - b \cdot b^*, a \cdot b^* + a^* \cdot b)$$

Remark 5.9 If you are puzzled by this definition of multiplication, then do not worry. In a few moments you will see that it gives rise to the notion of multiplication of complex numbers that you have seen before.

It is interesting to note that, unlike the integers and the rational numbers, the new number system $\mathbb{C}$ is *not* a collection of equivalence classes. Instead, $\mathbb{C}$ is the Euclidean plane equipped with some new algebraic operations.

EXAMPLE 5.13
Let $z = (3, -2)$ and $w = (4, 7)$ be two complex numbers. Then

$$z + w = (3, -2) + (4, 7) = (3 + 4, -2 + 7) = (7, 5).$$

Also

$$z \cdot w = (3, -2) \cdot (4, 7) = [3 \cdot 4 - (-2) \cdot 7, 3 \cdot 7 + 4 \cdot (-2)] = (26, 13) \quad \square$$

As usual, we ought to check that addition and multiplication are commutative, associative, that multiplication distributes over addition, and so forth. We shall leave these tasks as an exercise for the reader. Instead we develop some of the crucial properties of our new number system.

Theorem 5.12 *The following properties hold for the number system* $\mathbb{C}$.

(1) *The number* $1 \equiv (1, 0)$ *is the multiplicative identity:* $1 \cdot z = z$ *for any* $z \in \mathbb{C}$.

(2) *The number* $0 \equiv (0, 0)$ *is the additive identity:* $0 + z = z$ *for any* $z \in \mathbb{C}$.

(3) *Each complex number* $z = (x, y)$ *has an additive inverse* $-z = (-x, -y)$: *it holds that* $z + (-z) = 0$.

(4) *The number* $i \equiv (0, 1)$ *satisfies* $i \cdot i = (-1, 0) \equiv -1$; *in other words, i is a square root of* -1.

Proof: These are direct calculations, but it is important for us to work out these facts.

First, let $z = (x, y)$ be any complex number. Then

$$1 \cdot z = (1, 0) \cdot (x, y) = (1 \cdot x - 0 \cdot y, 1 \cdot y + x \cdot 0) = (x, y) = z$$

This proves the first assertion.

For the second, we have

$$0 + z = (0, 0) + (x, y) = (0 + x, 0 + y) = (x, y) = z$$

With z as above, set $-z = (-x, -y)$. Then

$$z + (-z) = (x, y) + (-x, -y) = [x + (-x), y + (-y)] = (0, 0) = 0$$

Finally, we calculate

$$i \cdot i = (0, 1) \cdot (0, 1) = (0 \cdot 0 - 1 \cdot 1, 0 \cdot 1 + 0 \cdot 1) = (-1, 0) = -1$$

Thus, as asserted, i is a square root of -1. $\qquad\qquad\square$

Proposition 5.4 *If $z \in \mathbb{C}, z \neq 0$, then there is a complex number w such that $z \cdot w = 1$.*

Proof: Write $z = (x, y)$ and set

$$w = \left(\frac{x}{x^2 + y^2}, \frac{-y}{x^2 + y^2} \right)$$

Since $z \neq 0$, this definition makes sense. Then it is straightforward to verify that $z \cdot w = 1$. $\qquad\qquad\square$

Thus every nonzero complex number has a multiplicative inverse. The other field axioms for $\mathbb{C}$ are easy to check. We conclude that the number system $\mathbb{C}$ forms a field. You will prove in the exercises that it is not possible to order this field. If α is a real number then we associate α with the complex number $(\alpha, 0)$. In this way, we can think of the real numbers as a *subset* of the complex numbers. In fact, the real field $\mathbb{R}$ is a *subfield* of the complex field $\mathbb{C}$. This means that if $\alpha, \beta \in \mathbb{R}$ and $(\alpha, 0), (\beta, 0)$ are the corresponding elements in $\mathbb{C}$ then $\alpha + \beta$ corresponds to $(\alpha + \beta, 0)$ and $\alpha \cdot \beta$ corresponds to $(\alpha, 0) \cdot (\beta, 0)$. These assertions are explored more thoroughly in the exercises.

With the remarks in the preceding paragraph we can sometimes ignore the distinction between the real numbers and the complex numbers. For example, we can write

$$5 \cdot i$$

and understand that it means $(5, 0) \cdot (0, 1) = (0, 5)$. Likewise, the expression

$$5 \cdot 1$$

can be interpreted as $5 \cdot 1 = 5$ or as $(5, 0) \cdot (1, 0) = (5, 0)$ without any danger of ambiguity or misunderstanding.

Theorem 5.13 *Every complex number can be written in the form $a + b \cdot i$, where a and b are real numbers. In fact, if $z = (x, y) \in \mathbb{C}$ then*

$$z = x + y \cdot i$$

Proof: With the identification of real numbers as a subfield of the complex numbers, we have that

$$x + y \cdot i = (x, 0) + (y, 0) \cdot (0, 1) = (x, 0) + (0, y) = (x, y) = z$$

as claimed. $\square$

Now that we have constructed the complex number field, we will adhere to the usual custom of writing complex numbers as $z = a + b \cdot i$ or, more simply, $a + bi$. We call a the *real part* of z, denoted by $Re\, z$, and b the *imaginary part* of z, denoted $Im\, z$. In this notation, our algebraic operations become

$$(a + bi) + (a^* + b^*i) = (a + a^*) + (b + b^*)i$$

and

$$(a + bi) \cdot (a^* + b^*i) = (a \cdot a^* - b \cdot b^*) + (a \cdot b^* + a^* \cdot b)i$$

If $z = a + bi$ is a complex number then we define its *complex conjugate* to be the number $\bar{z} = a - bi$. We record some elementary facts about the complex conjugate:

Proposition 5.5 *If z, w are complex numbers then*

(1) $\overline{z + w} = \bar{z} + \bar{w}$;

(2) $\overline{z \cdot w} = \bar{z} \cdot \bar{w}$;

(3) $z + \bar{z} = 2 \cdot Re\, z$;

(4) $z - \bar{z} = 2 \cdot i \cdot Im\, z$;

(5) $z \cdot \bar{z} \geq 0$, *with equality holding if and only if $z = 0$.*

Proof: Write $z = a + bi$, $w = c + di$. Then

$$\overline{z + w} = \overline{(a + c) + (b + d)i}$$
$$= (a + c) - (b + d)i$$
$$= (a - bi) + (c - di)$$
$$= \bar{z} + \bar{w}$$

This proves (1). Assertions (2), (3), (4) are proved similarly. For (5), notice that

$$z \cdot \bar{z} = (a + bi) \cdot (a - bi) = a^2 + b^2 \geq 0$$

Clearly equality holds if and only if $a = b = 0$. $\square$

The expression $|z|$ is defined to be the nonnegative square root of $z \cdot \bar{z}$. In other words

$$|z| = \sqrt{z \cdot \bar{z}} = \sqrt{(x + iy) \cdot (x - iy)} = \sqrt{x^2 + y^2}$$

It is called the *modulus* of z and plays the same role for the complex field that absolute value plays for the real field: the modulus of z measures the distance of z to the origin.

The modulus has the following properties.

Proposition 5.6 *If $z, w \in \mathbb{C}$ then*

1. $|z| = |\bar{z}|$;
2. $|z \cdot w| = |z| \cdot |w|$;
3. $|\text{Re } z| \leq |z|$, $|\text{Im } z| \leq |z|$;
4. $|z + w| \leq |z| + |w|$.

Proof: Write $z = a + bi$, $w = c + di$. Then (1), (2), and (3) are immediate. For (4) we calculate that

$$\begin{aligned}
|z + w|^2 &= (z + w) \cdot (\overline{z + w}) \\
&= z \cdot \bar{z} + z \cdot \bar{w} + w \cdot \bar{z} + w \cdot \bar{w} \\
&= |z|^2 + 2\text{Re}\,(z \cdot \bar{w}) + |w|^2 \\
&\leq |z|^2 + 2|z \cdot \bar{w}| + |w|^2 \\
&= |z|^2 + 2|z| \cdot |w| + |w|^2 \\
&= (|z| + |w|)^2
\end{aligned}$$

Taking square roots proves (4). $\square$

Observe that if z is real then $z = a + 0i$ and the modulus of z equals the absolute value of a. Likewise, if $z = 0 + bi$ is pure imaginary then the modulus of z equals the absolute value of b. In particular, the fourth part of the Proposition 5.6 reduces, in the real case, to the triangle inequality

$$|x + y| \leq |x| + |y|$$

5.8 The Quaternions, the Cayley Numbers, and Beyond

Now we shall discuss a number system that you may have never encountered before. It is called the system of *quaternions*. Our description will be an informal one.

Imagine $\mathbb{R}^4 \equiv \mathbb{R} \times \mathbb{R} \times \mathbb{R} \times \mathbb{R}$ equipped with the following operations: set $\mathbf{i} = (0, 1, 0, 0)$, $\mathbf{j} = (0, 0, 1, 0)$, $\mathbf{k} = (0, 0, 0, 1)$. Denote the 4-tuple $(1, 0, 0, 0)$ by $\mathbf{1}$. Define the multiplication laws

$$\mathbf{i} \cdot \mathbf{i} = -\mathbf{1} \qquad \mathbf{j} \cdot \mathbf{j} = -\mathbf{1} \qquad \mathbf{k} \cdot \mathbf{k} = -\mathbf{1}$$

and

$$\mathbf{i} \cdot \mathbf{j} = \mathbf{k} \qquad \mathbf{j} \cdot \mathbf{k} = \mathbf{i} \qquad \mathbf{k} \cdot \mathbf{i} = \mathbf{j}$$

and

$$\mathbf{j} \cdot \mathbf{i} = -\mathbf{k} \qquad \mathbf{k} \cdot \mathbf{j} = -\mathbf{i} \qquad \mathbf{i} \cdot \mathbf{k} = -\mathbf{j}$$

Of course the element $\mathbf{1}$ multiplied times any 4-tuple z is declared to be equal to z. In particular, $\mathbf{1} \cdot \mathbf{1} = \mathbf{1}$.

Finally, if $z = (z_1, z_2, z_3, z_4)$ and $w = (w_1, w_2, w_3, w_4)$ are 4-tuples then we write

$$z = z_1 \cdot \mathbf{1} + z_2 \mathbf{i} + z_3 \mathbf{j} + z_4 \mathbf{k}$$

and

$$w = w_1 \cdot \mathbf{1} + w_2 \mathbf{i} + w_3 \mathbf{j} + w_4 \mathbf{k}$$

Then $z \cdot w$ is defined by using the (obvious) distributive law and the rules already specified. For example,

$$(2, 0, 1, 3) \cdot (-4, 1, 0, 1) = [2 \cdot \mathbf{1} + \mathbf{j} + 3\mathbf{k}] \cdot [-4 \cdot \mathbf{1} + \mathbf{i} + \mathbf{k}]$$
$$= (2 \cdot (-4)) \cdot \mathbf{1} + (2\mathbf{i}) + (2\mathbf{k})$$
$$+ (\mathbf{j} \cdot (-4)) + (\mathbf{j} \cdot \mathbf{i}) + (\mathbf{j} \cdot \mathbf{k})$$
$$+ (3\mathbf{k} \cdot (-4)) + (3\mathbf{k} \cdot \mathbf{i}) + (3\mathbf{k} \cdot \mathbf{k})$$

$$= -8 \cdot \mathbf{1} + 2\mathbf{i} + 2\mathbf{k} - 4\mathbf{j} - \mathbf{k} + \mathbf{i} - 12\mathbf{k} + 3\mathbf{j} - 3 \cdot \mathbf{1}$$

$$= -11 \cdot \mathbf{1} + 3\mathbf{i} - \mathbf{j} - 11\mathbf{k}$$

$$= (-11, 3, -1, -11)$$

Addition of two quaternions is simply performed componentwise: if $z = (z_1, z_2, z_3, z_4)$ and $w = (w_1, w_2, w_3, w_4)$ then

$$z + w = (z_1 + w_1, z_2 + w_2, z_3 + w_3, z_4 + w_4)$$

Verify for yourself that the additive identity in the quaternions is $(0, 0, 0, 0)$. The multiplicative identity is $\mathbf{1} = (1, 0, 0, 0)$.

In fact it can be checked that each nonzero element of the quaternions has a unique two-sided multiplicative inverse. However, since multiplication is not commutative, the quaternions do not form a field; instead the algebraic structure is called a *division ring*.

It is also possible to give $\mathbb{R}^8$ an additive and a multiplicative structure. The multiplication operation is both noncommutative and nonassociative. The resulting eight-dimensional algebraic object is called the *Cayley numbers*. We shall not present the details here. It is one of the great theorems of twentieth century mathematics (see [ADA], [BOM]) that $\mathbb{R}^1$, $\mathbb{R}^2$, $\mathbb{R}^4$, and $\mathbb{R}^8$ are the only Euclidean spaces that can be equipped with compatible addition and multiplication operations in a natural way (so that the algebraic operations are smooth functions of the coordinates).

The quaternions and Cayley numbers are used in mathematical physics, in the representation theory of groups, and in algebraic topology. The cayley numbers are included in the ROM of every cell phone as part of the system of encoding messages.

Exercises

1. Let S be a set and let $p : S \times S \to S$ be a binary operation. If $T \subset S$ then we say that T is *closed* under p if $p : T \times T \to T$. (As an example, let $S = \mathbb{Z}$ and T be the even integers and p be ordinary addition.) Under which arithmetic operations $+, -, \cdot, \div$ is the set $\mathbb{Q}$ closed? Under which arithmetic operations $+, -, \cdot, \div$ is the set $\mathbb{R} \setminus \mathbb{Q}$ closed?

2. Let q be a rational number. Construct a sequence $\{x_j\}$ of irrational numbers such that $x_j \to q$. This means that, for each $\varepsilon > 0$, there is a positive integer K such that if $j > K$ then $|x_j - q| < \varepsilon$.

3. Let S be a set of real numbers with the property that whenever $x, y \in S$ and $x < t < y$ then $t \in S$. Can you give a simple description of the set S?

4. Explain why every nonzero complex number $\beta \in \mathbb{C}$ has two distinct square roots in $\mathbb{C}$.

5. An Argand diagram is a device for sketching a complex number in the plane. If $x + iy$ is a complex number then we depict it in the cartesian plane as the point (x, y). Sketch the complex numbers $3 - 2i$, $4i + 7$, $\pi i + e$, $-6 - i$.

6. The complex number $1 = 1 + 0i$ has three cube roots. Use any means to find them, and sketch them on an Argand diagram.

7. Let p be a polynomial and assume that $\alpha \in \mathbb{C}$ is a root of p. Prove that $(z - \alpha)$ evenly divides $p(z)$ with no remainder.

8. Determine whether $\sqrt{2} + \sqrt{3}$ is rational or irrational.

9. Find all square roots in the quaternions of the number $1 + \mathbf{i} + \mathbf{j}$.

10. Explain why subtraction in the integers is well defined.

11. Explain why multiplication in the integers is well defined.

12. Explain why division makes no sense in the integers.

13. Prove that addition in the integers is commutative.

14. Show that addition of rational numbers is well defined.

15. Show that the complex numbers cannot be ordered as a field.

16. Follow the outline in the text to show that the real numbers form a subfield of the complex numbers.

CHAPTER 6

Counting Arguments

Although everybody knows how to count in the sense of

1 2 3 4 5 ...

the fact is that counting is one of the most important techniques of modern mathematics. And counting can be quite sophisticated. Imagine counting the number of ways that one can get a straight flush in a hand of seven-card-stud poker. This is by no means a trivial counting problem.

In the present chapter we shall learn some important counting techniques that can be used to attack a variety of problems.

6.1 The Pigeonhole Principle

Also known as the *Dirichletscher Schubfachschluss* ("Dirichlet's drawer-shutting principle"), this is one of the key ideas in all of counting theory. And the idea is simplicity itself:

Pigeonhole principle: Let k be a positive integer. Imagine that you are delivering $k + 1$ letters to k mailboxes. Then it must be that some mailbox will receive two letters.

Obvious? Well, if each mailbox only received 0 or 1 letter then the total number of letters could not be more than

$$\underbrace{1 + 1 + \cdots + 1}_{k \text{ times}}$$

So there could not be more than k letters, and that is a contradiction (since we assumed that we had $k + 1$ letters). There are many other ways to verify this significant principle of counting.

Let us now illustrate the pigeonhole principle with some incisive examples.

EXAMPLE 6.1
Joe needs to get up at 4:00 a.m. each morning to go to work. He needs to get a pair of matching socks from his drawer without turning on the light and disturbing his wife. He knows that there are socks of three different colors, unpaired and randomly distributed, in the drawer. How many socks should he grab so that he can be sure to have a pair of the same color?

Solution: Call the sock colors red, green, and yellow. If Joe grabs three socks then one could be red, one could be green, and one could be yellow. So that will not do.

If he grabs four socks, then imagine that each sock is a "letter" and that there are three mailboxes labeled red, green, and yellow. He sticks each sock into the mailbox corresponding to its color. Well, there are three mailboxes and four letters, so some mailbox must end up with two letters. That means that two of the socks have the same color.

The answer is that Joe should grab four socks. □

EXAMPLE 6.2
There are 50 people in a room. Let us verify that there are two people in the room who have the same number of acquaintances in the room.

Solution: We consider two cases:

Case 1: There is one and only one person named Mary who has no acquaintances. Thus everyone else has 1 or 2 or ... or 48 acquaintances (nobody could have 49 acquaintances because then he/she would be acquainted with Mary—impossible!). Thus we have 49 people (these are the "letters"), each with somewhere between 1 and 48 acquaintances (these are the "mailboxes"). Thus some mailbox has two letters, meaning that two different people have the same number of acquaintances.

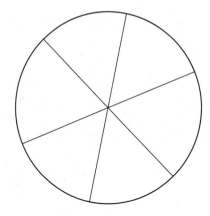

Figure 6.1   The dartboard divided into six regions.

Case 2: In this case we assume that there is no Mary. So *every one* of the 50 people has 1 or more acquaintances. Thus everyone has either 1 or 2 or ... or 49 acquaintances. That gives 50 letters going into 49 mailboxes. Some mailbox must contain two letters, meaning that two people have the same number of acquaintances.   □

EXAMPLE 6.3
Suppose that a standard dartboard has radius 10 inches. We throw seven darts at the dartboard. Why is it true that two of the darts will be distance at most 10 inches apart?

Solution: Examine Fig. 6.1. It shows the dartboard divided into six regions. But there are seven darts. By the pigeonhole principle, two of the darts must land in the same region. Refer to Fig. 6.2. Since the dartboard has radius 10 inches, no two points of the region are more distant than 10 inches. So these two darts satisfy our criterion.   □

EXAMPLE 6.4
Verify that, at any given moment in New York City, there are two people with the same number of hairs on their heads.

Solution: Any person has at most 900 hairs per square inch on his/her dome. No head has diameter greater than 20 inches. It is easy to estimate that the cranium has surface area at most $4\pi r^2$ (that is, the surface area of a sphere of radius r) hence at most 5000 square inches. So nobody has more that 5 million hairs on his/her head.

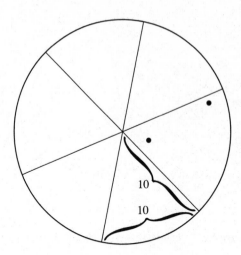

Figure 6.2 Two darts in the same region.

But New York City has about 9 million people. Thus we have 9 million letters with at most 5 million mailboxes. It follows that some mailbox will contain two letters, so that two people will have the same number of hairs. □

You Try It: Select 55 integers at random between 1 and 100 inclusive. Demonstrate that two of these selected numbers will differ by exactly 9.

You Try It: There are 9 people seated in a row of 12 chairs. Show that three consecutive chairs must be occupied.

6.2 Orders and Permutations

Suppose that we have n objects, as in Fig. 6.3. In how many different orders can they be presented? The technical language for this question is, "How many different permutations are there of n objects?" Here a *permutation* is simply a reordering of the objects.

Figure 6.3 An array of n objects.

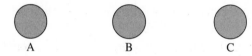

Figure 6.4   An array of three objects.

As a simple example, take n to be 3. Refer to Fig. 6.4. In how many different orders can we present these objects? Fig. 6.5 shows the different ways.

We see that there are a total of six different ways to permute three objects. Contrast this situation with two objects—there are only two different ways to permute two objects ($\{A, B\}$ or $\{B, A\}$). In fact the number of permutations of n objects grows rather rapidly with n.

Now let us return to n objects. In how many different ways can we order them? Look at the first position. We can put any of the n objects in that first position, so there are n possibilities for the first position. Next go to the second position.

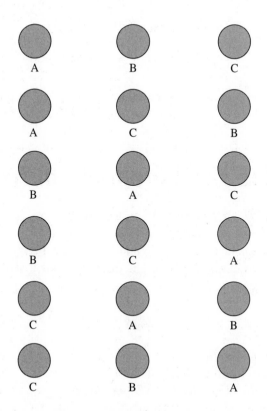

Figure 6.5   All the different orders for three objects.

One object is used up, so there are $n - 1$ objects remaining. *Any one* of those $n - 1$ objects can go into the second position.

Now examine the third position. There are $n - 2$ objects remaining (since two of them are used up). So any of those $n - 2$ can go in the third position. And so forth. We see that there are three possibilities for the $(n - 2)$th position, two possibilities for the $(n - 1)$th position, and one possibility for the nth position.

To count the total number of possible orderings, we multiply together all these counts:

$$\text{number of permutations of } n \text{ objects} = n \cdot (n - 1) \cdot (n - 2) \cdots \cdot 3 \cdot 2 \cdot 1$$

This last expression is so important, and so pervasive, in mathematics that we give it a name. It is called *n factorial*, and is written $n!$.

EXAMPLE 6.5
In how many different ways can we order five objects?

Solution: The number of permutations of five objects is

$$5! = 5 \cdot 4 \cdot 3 \cdot 2 \cdot 1 = 120$$

There are 120 different permutations of 5 objects. □

6.3 Choosing and the Binomial Coefficients

Suppose that you have n objects and you are going to choose k of them (where $0 \leq k \leq n$). In how many different ways can you do this?

Just to illustrate the idea, take $n = 3$ and $k = 2$—see Fig. 6.6. For convenience we have labeled the objects A, B, C. The different ways that we can choose two from among the three are

$$\{A, B\} \qquad \{A, C\} \qquad \{B, C\}$$

There are no other possibilities.

It is of interest to try to analyze this problem in general. So imagine now that we have n objects as shown in Fig. 6.7. We are going to select k of them, with $0 \leq k \leq n$. We may note in advance that there is only 1 way to select 0 objects from among n—you just do not select any! So we may assume that k is at least 1.

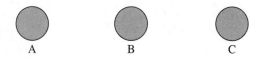

Figure 6.6   Three objects from which to choose two.

We select the first object. There are n different ways to do this—either you select the first one a_1, or the second one a_2, or ... or the last one a_n.

Now let us think about selecting the second object. One object has already been selected, so there are $n - 1$ objects left. And we may select any one of them with no restriction. So there are $n - 1$ ways to select the second object. See Fig. 6.8.

Now the pattern emerges. When we go to select the third object, there are $n - 2$ choices remaining. So there are $n - 2$ ways to select the third object.

And on it goes. There are $n - 3$ ways to select the fourth object, $n - 4$ ways to select the fifth object, ..., $n - k + 1$ ways to select the kth object.

Altogether then we have

$$n \cdot (n - 1) \cdot (n - 2) \cdots (n - k + 1)$$

ways to select k objects from among n total. But we have overlooked an important fact—refer to the last section. We could have selected those k objects in any of $k!$ different orders. So in fact the number of ways to select k objects from among n is

$$\frac{n \cdot (n - 1) \cdot (n - 2) \cdots (n - k + 1)}{k!}$$

This is a very common expression in mathematics, and we give it the name n *choose* k. We denote this quantity by $\binom{n}{k}$. In fact it is customary to write it as

$$\binom{n}{k} = \frac{n!}{(n - k)!k!}$$

EXAMPLE 6.6

In how many different ways can we choose two objects from among five?

Figure 6.7   Selecting k objects from among n.

Figure 6.8 One object selected and $n - 1$ remaining.

Solution: The answer is immediate:

$$\binom{5}{2} = \frac{5!}{(5-2)!2!} = \frac{120}{6 \cdot 2} = 10$$

In fact if the five objects are a, b, c, d, and e then the five possible choices of two are:

$$\{a, b\} \quad \{a, c\} \quad \{a, d\} \quad \{a, e\} \quad \{b, c\}$$
$$\{b, d\} \quad \{b, e\} \quad \{c, d\} \quad \{c, e\} \quad \{d, e\} \qquad \square$$

EXAMPLE 6.7
How many different 5-card poker hands are there in a standard 52-card deck of cards?

Solution: The answer is

$$\binom{52}{5} = \frac{52!}{(52-5)!5!} = 2,598,960 \qquad \square$$

EXAMPLE 6.8
In a five-card poker hand, what are the chances of having three-of-a-kind?

Solution: For the three matching cards, the first card that we select can be anything. So there are 52 possibilities. But the next one must match it. So there are only three choices for the second card (remember that there are four of each kind of card in the deck). And for the third card there are then just two possibilities. The other two cards can be completely random, so there are 49 and 48 possibilities for those two cards. Thus the total number of ways to have a hand with three-of-a-kind is

$$\frac{52 \cdot 3 \cdot 2 \cdot 49 \cdot 48}{6} = 122,304$$

Note the little surprise: we have divided by 6 since we must divide out all the different possible orders of the same set of three matching cards; the number of permutations of three objects is $3! = 6$.

Now the *odds* of getting three-of-a-kind is the ratio

$$\frac{122304}{2598960} = 0.047058824$$

(using the count of all possible hands from the previous example) or slightly less than 1 in 20. □

You Try It: What are the odds of getting two pair in a standard five-card poker hand?

You Try It: What are the odds of getting four-of-a-kind in a standard five-card poker hand?

You Try It: You have a pot of beads. The beads are all identical in size and shape, but come in 2 different colors. You wish to make a beaded necklace consisting of 10 beads. How many different necklaces could you make? Of course *the order* of the beads will be important: beads in the order black-white-black-white does *not* give the same necklace as black-black-white-white. Note also that two necklaces are *equivalent*, and count as just one necklace, if a rotation of one gives the other. [After you have solved this problem, try replacing "10" with n and "2" with k and solve it again.]

6.4 Other Counting Arguments

It will be useful for this section and the material that follows to have standard mathematical summation notation at our disposal. The expression

$$\sum_{j=1}^{N} a_j \tag{6.1}$$

is used to denote

$$a_1 + a_2 + \cdots + a_N$$

Observe that the symbol $\sum$ is the Greek letter sigma, which is a cognate of our roman S. We read the Eq. (6.1) as meaning that we sum a_j as the index j ranges

from 1 to N. We will also write

$$\sum_{j=5}^{N} b_j$$

to indicate that b_j should be summed from 5 to N. In fact we allow any integral lower index M and any upper index N; the custom is that M should be less than or equal to N.

Sometimes we also write

$$\sum_{j=1}^{\infty} c_j$$

to indicate that we are summing infinitely many terms. This idea will be explored in Chap. 13.

EXAMPLE 6.9
Draw a planar grid that is 31 squares wide and 17 squares high. How many different nontrivial rectangles can be drawn, using the lines of the grid to determine the boundaries? See Fig. 6.9. (Here "nontrivial" means that the rectangle has positive width and positive height.)

Solution: We need a cogent method for counting the rectangles. Note that any rectangle is uniquely specified by the location of its lower left-hand corner, its length, and its width.

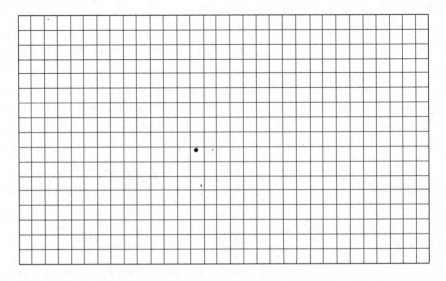

Figure 6.9 A 31 × 17 grid.

We think of the lower left point of the grid in Fig. 6.9 as the origin and locate points in the usual cartesian fashion (with the side of a square in the grid being a unit).

How many rectangles can have their lower left corner at the origin? Well, there are 31 possible widths, from 1 to 31, and 17 possible heights, from 1 to 17. That is a total of $31 \times 17 = 527$ possible rectangles with lower left corner at 0.

This is a good start, but extrapolation from this beginning will constitute an awfully tedious manner of enumerating *all* the possible rectangles. We will now be a bit more analytical. Suppose that we consider rectangles with lower left corner at (j, k), with $0 \le j \le 30$ and $0 \le k \le 16$. There are $(31 - j)$ possible widths for such rectangles and $(17 - k)$ possible heights. Thus, altogether, there are $(31 - j) \times (17 - k)$ rectangles with lower left corner at (j, k). Given the range of j and k, the total number of all possible nontrivial rectangles is

$$S = \sum_{j=0}^{30} \sum_{k=0}^{16} (31 - j) \times (17 - k)$$

It is useful to expand the summands using the distributive law and then to regroup. We have

$$S = \sum_{j=0}^{30} \sum_{k=0}^{16} \left[527 - 17j - 31k + jk \right]$$

This, in turn, equals

$$\sum_{j=0}^{30} \sum_{k=0}^{16} 527 - \sum_{j=0}^{30} \sum_{k=0}^{16} 17j - \sum_{j=0}^{30} \sum_{k=0}^{16} 31k + \sum_{j=0}^{30} \sum_{k=0}^{16} jk = [527 \cdot 31 \cdot 17]$$

$$- 17 \cdot 17 \cdot \sum_{j=0}^{30} j - 31 \cdot 31 \cdot \sum_{k=0}^{16} k + \left[\sum_{j=0}^{30} j \right] \cdot \left[\sum_{k=0}^{16} k \right]$$

Now we may use Gauss's formula from Sec. 2.4 to evaluate each of the sums in the last formula. We obtain

$$S = 277729 - 289 \cdot 465 - 961 \cdot 136 + 465 \cdot 136$$

$$= 277729 - 134385 - 130696 + 63240$$

$$= 75888 \qquad \square$$

In Chap. 1 we saw how to use the method of induction to calculate certain finite sums, such as $1 + 2 + 3 + \cdots k$, in closed form. Now we look at another type of sum, known as a *geometric sum*.

A geometric sum is a sum of powers of a fixed number. For example,

$$1 + 3 + 3^2 + 3^3 + \cdots + 3^{10}$$

is a finite geometric sum. Also

$$1 + \frac{1}{2} + \left(\frac{1}{2}\right)^2 + \left(\frac{1}{2}\right)^3 + \cdots$$

is an infinite geometric sum. Notice in each case that the first term of the sum is 1, and that is the zeroeth power of the fixed number λ (in the first instance 3 and in the second instance $1/2$).

EXAMPLE 6.10
Let λ be a real number and k a positive integer. Calculate the sum of the geometric series

$$S = 1 + \lambda + \lambda^2 + \cdots + \lambda^k$$

Solution: The key is to note that multiplying S by λ does not change it very much. Indeed

$$\lambda S = \lambda + \lambda^2 + \lambda^3 + \cdots + \lambda^{k+1}$$

The sums S and λS differ only in the presence of 1 in the first of these and the presence of λ^{k+1} in the second. In other words

$$S - 1 = \lambda S - \lambda^{k+1}$$

That is,

$$(\lambda - 1)S = \lambda^{k+1} - 1$$

We finally write this as

$$S = \frac{\lambda^{k+1} - 1}{\lambda - 1} \qquad \square$$

An example of what the solution of the last problem tells us is as follows: Suppose that we want to know explicitly the value of the sum $S = 1 + (1/3) + (1/3)^2 + \cdots + (1/3)^{100}$. It would be quite tedious to add all these numbers up by hand (or even using a calculator). But this question fits the paradigm of the geometric series, with $\lambda = 1/3$ and $k = 100$. Thus

$$S = \frac{(1/3)^{101} - 1}{(1/3) - 1} = \frac{3}{2} \cdot \left(1 - \left[\frac{1}{3}\right]^{101}\right)$$

Now a calculator may be used to see that the value of this last expression is about $1.5 - 9.702 \cdot 10^{-49}$.

Sometimes it is convenient, when $-1 < \lambda < 1$, to reason as follows: for $k \in \{1, 2, 3, \ldots\}$ we set

$$S_k = 1 + \lambda + \lambda^2 + \cdots \lambda^k$$

We know, from the last example, that

$$S_k = \frac{1 - \lambda^{k+1}}{1 - \lambda} \qquad (6.2)$$

Now we could ask what happens if, instead of adding just finitely many powers of λ, we add *all* powers of λ. This would correspond (in a sense that you will learn about more precisely when we get to Chaps. 12 and 13) to letting k tend to infinity in Eq. (6.2).

The result is that the sum $S = 1 + \lambda + \lambda^2 + \cdots$ of *all* nonnegative powers of λ is obtained by asking what happens to the right-hand side of Eq. (6.2) when k becomes large without bound. Since $|\lambda| < 1$, it is plausible that λ^{k+1} becomes smaller and smaller, in fact tends to 0, as k increases without bound. In other words, $S_k \to 1/(1 - \lambda)$. We write

$$\sum_{j=0}^{\infty} \lambda^j = \frac{1}{1 - \lambda} \qquad (6.3)$$

This is a variant of the standard mathematical notation for summation. The symbol $\sum$ denotes the summation process. The lower limit means that we begin our summing with the exponent j equaling 0 and the upper limit having no bound (in other words, we sum *all powers* of λ).

Here is an illustrative example: what is the sum

$$1 + \frac{1}{2} + \left(\frac{1}{2}\right)^2 + \left(\frac{1}{2}\right)^3 + \cdots$$

equal to? Draw a picture of the interval $[0, 2]$ on your scratch pad. The sum of the first two terms is $3/2$. Add one additional term and you cover half the remaining distance to 2. Add the fourth term and you again cover half the remaining distance to 2. In fact each additional term repeats this key property. It is plausible to suppose that the *entire sum* equals 2.

In fact our new formula makes this supposition concrete:

$$\sum_{j=0}^{\infty}\left(\frac{1}{2}\right)^{j} = \frac{1}{1-(1/2)} = 2$$

6.5 Generating Functions

Now we shall learn the powerful technique of generating functions. In fact we shall build on what has gone before. We will use our new ideas about geometric series, in a very simple form, in the next problem.

EXAMPLE 6.11
The Fibonacci sequence is famous in mathematics, indeed in all of science. The Fibonacci sequence describes the spacing of leaves on a vine, the turns of a conch shell, and many other natural phenomena.

The sequence is formed in the following way: the first two terms are each equal to 1. The next term is obtained by adding the preceding two: thus the third term equals 2. The next term (the fourth) is obtained by adding the preceding two: $1 + 2 = 3$. The next term is obtained by adding the preceding two: $2 + 3 = 5$. In fact the first 10 terms of the Fibonacci sequence are

$$1, 1, 2, 3, 5, 8, 13, 21, 34, 55$$

We denote the jth term of the Fibonacci sequence by a_j. Thus

$$a_0 = 1 \quad a_1 = 1 \quad a_2 = 2 \quad a_3 = 3 \quad a_4 = 5$$

and so forth.

Show that the following formula for the Fibonacci sequence is valid:

$$a_j = \frac{\left(\frac{1+\sqrt{5}}{2}\right)^{j} - \left(\frac{1-\sqrt{5}}{2}\right)^{j}}{\sqrt{5}}$$

Solution: We shall use the method of *generating functions*, a powerful technique that is used throughout the mathematical sciences.

We write $F(x) = a_0 + a_1 x + a_2 x^2 + \cdots$. Here the a_j's are the terms of the Fibonacci sequence and the letter x denotes an unspecified variable. What is curious here is that we do not care about what x is. It is simply an unspecified variable. We *shall not* solve for x. We intend to manipulate the function F in such a fashion that we will be able to solve for the coefficients a_j. Just think of $F(x)$ as a polynomial with a *lot* of coefficients.

Notice that

$$xF(x) = a_0 x + a_1 x^2 + a_2 x^3 + a_3 x^4 + \cdots$$

and

$$x^2 F(x) = a_0 x^2 + a_1 x^3 + a_2 x^4 + a_3 x^5 + \cdots$$

Thus, grouping like powers of x, we see that

$$
\begin{aligned}
F(x) &- xF(x) - x^2 F(x) \\
&= a_0 + (a_1 - a_0)x + (a_2 - a_1 - a_0)x^2 \\
&\quad + (a_3 - a_2 - a_1)x^3 + (a_4 - a_3 - a_2)x^4 + \cdots
\end{aligned}
$$

But the basic property that defines the Fibonacci sequence is that $a_2 - a_1 - a_0 = 0$, $a_3 - a_2 - a_1 = 0$, and so on. Thus our equation simplifies drastically to

$$F(x) - xF(x) - x^2 F(x) = a_0 + (a_1 - a_0)x$$

We also know that $a_0 = a_1 = 1$. Thus the equation becomes

$$(1 - x - x^2)F(x) = 1$$

or

$$F(x) = \frac{1}{1 - x - x^2} \tag{6.4}$$

It is convenient to factor the denominator as follows:

$$F(x) = \frac{1}{\left[1 - \frac{-2}{1-\sqrt{5}}x\right] \cdot \left[1 - \frac{-2}{1+\sqrt{5}}x\right]}$$

[just simplify the right-hand side to see that it equals Eq. (6.4)].

A little more algebraic manipulation yields that

$$F(x) = \frac{5 + \sqrt{5}}{10}\left[\frac{1}{1 + \frac{2}{1-\sqrt{5}}x}\right] + \frac{5 - \sqrt{5}}{10}\left[\frac{1}{1 + \frac{2}{1+\sqrt{5}}x}\right]$$

Now we want to apply the formula in Eq. (6.2) from Sec. 6.4 to each of the fractions in brackets ([ ]). For the first fraction, we think of $-\frac{2}{1-\sqrt{5}}x$ as λ. Thus the first expression in brackets equals

$$\sum_{j=0}^{\infty}\left(-\frac{2}{1 - \sqrt{5}}x\right)^{j}$$

Likewise the second sum equals

$$\sum_{j=0}^{\infty}\left(-\frac{2}{1 + \sqrt{5}}x\right)^{j}$$

All told, we find that

$$F(x) = \frac{5 + \sqrt{5}}{10}\sum_{j=0}^{\infty}\left(-\frac{2}{1 - \sqrt{5}}x\right)^{j} + \frac{5 - \sqrt{5}}{10}\sum_{j=0}^{\infty}\left(-\frac{2}{1 + \sqrt{5}}x\right)^{j}$$

Grouping terms with like powers of x, we finally conclude that

$$F(x) = \sum_{j=0}^{\infty}\left[\frac{5 + \sqrt{5}}{10}\left(-\frac{2}{1 - \sqrt{5}}\right)^{j} + \frac{5 - \sqrt{5}}{10}\left(-\frac{2}{1 + \sqrt{5}}\right)^{j}\right]x^{j}$$

But we began our solution of this problem with the formula

$$F(x) = a_0 + a_1 x + a_2 x^2 + \cdots$$

The two different formulas for $F(x)$ must agree. In particular, the coefficients of the different powers of x must match up. We conclude that

$$a_j = \frac{5 + \sqrt{5}}{10}\left(-\frac{2}{1 - \sqrt{5}}\right)^{j} + \frac{5 - \sqrt{5}}{10}\left(-\frac{2}{1 + \sqrt{5}}\right)^{j}$$

We rewrite

$$\frac{5+\sqrt{5}}{10} = \frac{1}{\sqrt{5}} \cdot \frac{1+\sqrt{5}}{2} \qquad \frac{5-\sqrt{5}}{10} = -\frac{1}{\sqrt{5}} \cdot \frac{1-\sqrt{5}}{2}$$

and

$$-\frac{2}{1-\sqrt{5}} = \frac{1+\sqrt{5}}{2} \qquad -\frac{2}{1+\sqrt{5}} = \frac{1-\sqrt{5}}{2}$$

Making these four substitutions into our formula for a_j, and doing a few algebraic simplifications, yields

$$a_j = \frac{\left(\frac{1+\sqrt{5}}{2}\right)^j - \left(\frac{1-\sqrt{5}}{2}\right)^j}{\sqrt{5}}$$

as desired.   □

Notice how, in this last example, we combined F, xF, and x^2F so that important cancellations would take place. That is how we used the special properties of the Fibonacci sequence. In other problems, such as those in the next section, you will need to use different combinations, with possibly different coefficients, that are tailored to each specific problem.

6.6 A Few Words About Recursion Relations

Let $\{a_j\}$ be a sequence, or a list of numbers. More explicitly, the sequence is

$$a_0, a_1, a_2, \ldots$$

where the list never stops. It is frequently the case that we will have a rule that tells us the value of the jth element of the list in terms of some of the previous elements. This situation is called a *recursion*. The method of generating functions can sometimes be used to good effect to solve recursions. We shall illustrate the idea here with some examples.

EXAMPLE 6.12
A sequence is defined by the rule $a_0 = 4$, $a_1 = -1$, and $a_j = -a_{j-1} + 2a_{j-2}$. Use the method of generating functions to find a formula for a_j.

Remark 6.1 Notice that the recursion rule says in particular that

$$a_2 = -a_1 + 2a_0$$

As a result, $a_2 = 9$. Similarly,

$$a_3 = -a_2 + 2a_1$$

It follows that $a_3 = -11$.

This is how recursions work.

Solution: We write $F(x) = a_0 + a_1x + a_2x^2 + \cdots$. Here the a_js are the terms of the unknown sequence and the letter x denotes an unspecified variable.

Notice that

$$xF(x) = a_0x + a_1x^2 + a_2x^3 + a_3x^4 + \cdots$$

and

$$x^2F(x) = a_0x^2 + a_1x^3 + a_2x^4 + a_3x^5 + \cdots$$

Thus, grouping like powers of x, we see that

$$F(x) + xF(x) - 2x^2F(x)$$
$$= a_0 + (a_1 + a_0)x + (a_2 + a_1 - 2a_0)x^2$$
$$+ (a_3 + a_2 - 2a_1)x^3 + \cdots$$

But the basic property that defines our sequence is that $a_2 = -a_1 + 2a_0$, $a_3 = -a_2 + 2a_1$, and so on. Thus our equation simplifies drastically to

$$F(x) + xF(x) - 2x^2F(x) = a_0 + (a_1 + a_0)x$$

We also know that $a_0 = 4$ and $a_1 = -1$. Thus the equation becomes

$$(1 + x - 2x^2)F(x) = 4 + 3x$$

or

$$F(x) = \frac{4 + 3x}{1 + x - 2x^2} \tag{6.5}$$

It is convenient to factor the denominator as follows:

$$F(x) = \frac{4 + 3x}{(-2x - 1) \cdot (x - 1)}$$

[just simplify the right-hand side to see that it equals Eq. (6.5)].
    A little more algebraic manipulation yields that

$$F(x) = \left(-\frac{5}{3}\right) \cdot \frac{1}{-2x - 1} + \left(-\frac{7}{3}\right) \cdot \frac{1}{x - 1}$$

$$= \left(\frac{5}{3}\right) \cdot \frac{1}{1 - (-2x)} + \left(\frac{7}{3}\right) \cdot \frac{1}{1 - x}$$

Now we want to apply the formula in Eq. (6.2) from Sec. 6.4 to each of the fractions here. For the first fraction, we think of $-2x$ as λ. Thus the first fractional expression equals

$$\sum_{j=0}^{\infty} (-2x)^j$$

Likewise the second fractional expression equals

$$\sum_{j=0}^{\infty} x^j$$

All told, we find that

$$F(x) = \frac{5}{3} \sum_{j=0}^{\infty} (-2x)^j + \frac{7}{3} \sum_{j=0}^{\infty} x^j$$

Grouping terms with like powers of x, we finally conclude that

$$F(x) = \sum_{j=0}^{\infty} \left[\frac{5}{3}(-2)^j + \frac{7}{3} \right] x^j$$

But we began our solution of this problem with the formula

$$F(x) = \sum_{j=0}^{\infty} a_j x^j$$

The two different formulas for $F(x)$ must agree. In particular, the coefficients of the different powers of x must match up. We conclude that

$$a_j = \frac{5}{3}(-2)^j + \frac{7}{3}$$

That is the solution to our recursion problem. $\square$

You Try It: A sequence is defined by the rule $a_0 = 2$, $a_1 = 1$, and $a_j = 3a_{j-1} - a_{j-2}$. Use the method of generating functions to find a formula for a_j.

You Try It: A sequence is defined by the rule $a_0 = 0$, $a_1 = -1$, and $a_j = 3a_{j-1} - 2a_{j-2}$. Use the method of generating functions to find a formula for a_j.

6.7 Probability

In Sec. 6.3 we have alluded to certain questions of probability theory. Now we take a moment to treat the subject a bit more formally and precisely.

If an event E has finitely many possible outcomes $o_1, o_2, \ldots, o_k$, then we assign a positive number p_j to each outcome o_j to indicate the likelihood that outcome will actually occur. We mandate in advance that

$$\sum_{j=1}^{k} p_j = 1$$

indicating that 1 describes the totality of all outcomes.

If we are flipping a coin (that is the event E), then there are two possible outcomes: heads and tails. Observation of many flips (or just common sense) teaches us that the two outcomes are equally likely. Therefore we assign the number $p_h = 1/2$ to the outcome "heads" and the number $p_t = 1/2$ to the outcome "tails." Notice that $1/2 + 1/2 = 1$, as we have mandated. We call p_h the *probability* that heads will occur and p_t the *probability* that tails will occur.

It is common to express probabilities in terms of percentage. We might also say (referring to the last paragraph) that there is a 50% probability that heads will occur and a 50% probability that tails will occur.

Now let us look at a different situation. I hold a red ball and a blue ball. So do you. Each of us will randomly put one of the balls into a box on the table. What are the probabilities of the different outcomes? The different possibilities are BB, RR, and RB (here R stands for "red" and B stands for "blue"). A naive analysis might cause one to guess that each of these three outcomes has the same probability.[1] But the naive analysis is wrong. Here is why:

- The outcome *BB* can only occur if each of us contributes a blue ball. So it can only happen one way.
- The outcome *RR* can only occur if each of us contributes a red ball. So it can only happen one way.
- The outcome *RB* can occur if I contribute a red ball and you contribute a blue, or if I contribute a blue ball and you contribute a red. So it can happen in two different ways.

Thus we see that the correct assignment of probabilities is $p_{BB} = 1/4$, $p_{RR} = 1/4$, $p_{RB} = 1/2$. We have chosen these numbers to meet the following criteria:

- The sum of all the probabilities should be 1.
- The probability of *BB* and *RR* should be equal.
- The probability of *RB* should be double that of *BB* or *RR*.

EXAMPLE 6.13

A girl flips a fair coin five times. What is the probability that precisely three of the flips will come up heads?

Solution: Each flip has two possible outcomes. So the total number of possible outcomes for five flips is

$$2 \cdot 2 \cdot 2 \cdot 2 \cdot 2 = 32$$

Now the number of ways that three head flips can occur is the number of ways that three objects can be chosen from five. This is

$$\binom{5}{3} = \frac{5!}{3! \cdot 2!} = 10$$

We conclude that the answer to the question is

$$p = \frac{10}{32} = 0.3125$$ □

---

[1]In fact this is what genetecists in the early twentieth century thought—where for them *B* is a dominant gene and *R* is a recessive gene. It was eminent mathematician G. H. Hardy who noted the error of their ways. The resulting published paper led to what is now called the *Hardy-Weinberg law* in genetics.

You Try It: Calculate the probability that the girl in the last example will get zero heads, one head, two heads, four heads, and five heads. Add up all these results (including the result for three heads from the last example). The answer should of course be 1, since these are all the possible outcomes.

EXAMPLE 6.14
Eight slips of paper with the letters $A, B, C, D, E\ F, G$, and H written on them are placed into a bin. The eight slips are drawn one by one from the bin. What is the probability that the first four to come out are A, C, E, and H (in *some* order)?

Solution: This problem is much less exciting than it sounds. After we choose the first four slips, it does not matter what we do. We could burn the others, or go drink coffee, or enroll in truck driving school. And the statement of the problem rules out the *order* in which the slips are drawn. Stripping away the language, we see that we are randomly selecting four objects from among eight. We want to know whether a particular four, in any order, will be the ones that we select.

The number of different ways to choose four objects from among eight is

$$\binom{8}{4} = \frac{8!}{4! \cdot 4!} = \frac{8 \cdot 7 \cdot 6 \cdot 5}{4 \cdot 3 \cdot 2 \cdot 1} = 70$$

Of these different subsets of four, only one will be the set $\{A, C, E, H\}$. Thus the probability of the first four slips being the ones that we want will be 1/70. $\square$

EXAMPLE 6.15
Suppose that you write 37 letters and then you address 37 envelopes to go with them. Closing your eyes, you randomly stuff one letter into each envelope. What is the probability that just one envelope contains the wrong letter?

Solution: Say that the envelopes are numbered 1–37 and the letters are numbered 1–37. If letters 1 through 36 go into envelopes 1–36 then what remains are letter 37 and envelope 37. So that last letter is *forced* to go into the correct envelope.

Of course there is nothing special about the numbering used in the last paragraph. It just helped us to make a simple point: it is impossible to have just one letter in the wrong envelope. If one letter is in the wrong envelope then at least two letters are in the wrong envelope.

Thus the answer to our problem is that the probability is zero. $\square$

You Try It: You have an urn with 100 black marbles and 100 white marbles. You close your eyes and grab five marbles at random. What is the probability that at least three of them are black?

EXAMPLE 6.16
Suppose that you have 37 envelopes and you address 37 letters to go with them. Closing your eyes, you randomly stuff one letter into each envelope. What is the probability that *precisely* two letters are in the wrong envelopes and all others in the correct envelope?

Solution: If just two letters are to be in the wrong envelope then they will have to be switched; for instance, letter 5 could go into envelope 19 and letter 19 into envelope 5. Thus the number of different ways that we can get just two letters in the wrong envelopes is just the same as the number of different ways that we can choose two letters from among 37. (All of the other 35 letters must go into their *correct* envelopes, so there is no choice involved for those 35.) This number is

$$N = \binom{37}{2} = \frac{37!}{2!35!} = \frac{37 \cdot 36}{2 \cdot 1} = 666$$

Now if we imagine the envelopes, in their correct order (numbers 1–37), lying in a row on the table, then a random distribution of letters among the envelopes just corresponds to a random ordering of the letters. Thus the number of different possible ways to distribute 37 letters among 37 envelopes is 37! (a very large number). In conclusion, the probability that all letters but two will be in the correct envelopes is

$$P = \frac{666}{37!} \approx 4.86 \cdot 10^{-41}$$

□

6.8 Pascal's Triangle

The idea of what we now call *Pascal's triangle* actually goes back to Yanghui in about the twelfth century in China (and the Chinese call the object *Yanghui's triangle*). But it was Blaise Pascal (1623–1662) who really developed the concept and showed its importance and context in modern mathematics.

Pascal's triangle is a triangle formed according to the following precept (see Fig. 6.10).

The rule for forming Pascal's triangle is this:

- A 1 goes at the top vertex.
- Each term in each subsequent row is formed by adding together the two numbers that are to the upper left and upper right of the given term.

$$
\begin{array}{ccccccccccc}
 & & & & & 1 & & & & & \\
 & & & & 1 & & 1 & & & & \\
 & & & 1 & & 2 & & 1 & & & \\
 & & 1 & & 3 & & 3 & & 1 & & \\
 & 1 & & 4 & & 6 & & 4 & & 1 & \\
1 & & 5 & & 10 & & 10 & & 5 & & 1 \\
 & & & & & \cdots & & & & &
\end{array}
$$

Figure 6.10 Pascal's triangle.

It is convenient in our discussion to refer to the very top row of Pascal's triangle (with a single digit 1 in it) as the zeroth row. The next row is the first row. And so forth. So the zeroth row has one element, the first row has two elements, the second row has three elements, and so forth.

Thus, in the first row, the leftmost term has nothing to its upper left and a 1 to its upper right. The sum of these is 1. So the leftmost term in the first row is 1. Likewise the rightmost term in the first row is 1.

For the second row, the leftmost term has nothing to its upper left and a 1 to its upper right. So this new term is 1. Likewise the rightmost term in the second row is 1. But the middle term in the second row has a 1 to its upper left and a 1 to its upper right. Therefore this middle term equals $1 + 1 = 2$. That is what we see in Pascal's triangle.

For the third row, we see as usual that the leftmost term and the rightmost terms are both 1 (in fact this property holds in all rows). But the second term in the third row has a 1 to its upper left and a 2 to its upper right. Therefore this second terms is equal to 3. Likewise the third term is equal to 3.

The rest of Pascal's triangle is calculated similarly. The triangle is obviously symmetric from left to right, about a vertical axis through the upper vertex. The kth row has $k + 1$ terms. What is the significance of these numbers?

One obvious significance is the relation of the triangle to the celebrated binomial theorem. Consider the quantity

$$
(a + b)^k = a^k + \frac{k}{1}a^{k-1}b + \frac{k}{2}a^{k-2}b^2 + \frac{k}{3}a^{k-3}b^3 + \cdots
$$

$$
+ \frac{k}{k-3}a^3 b^{k-3} + \frac{k}{k-2}a^2 b^{k-2} + \frac{k}{k-1}ab^{k-1} + \frac{k}{k}b^k
$$

Now we will examine this important formula in the first several specific instances:

k = 0: $(a + b)^0 = 1$

k = 1: $(a + b)^1 = a + b$

k = 2: $(a + b)^2 = a^2 + 2ab + b^2$

k = 3: $(a + b)^3 = a^3 + 3a^2b + 3ab^2 + b^3$

k = 4: $(a + b)^4 = a^4 + 4a^3b + 6a^2b^2 + 4ab^3 + b^4$

k = 5: $(a + b)^5 = a^5 + 5a^4b + 10a^3b^2 + 10a^2b^3 + 5ab^4 + b^5$

We see that the coefficients that occur for **k = 0** are just the same as the zeroeth row of Pascal's triangle: namely, a single digit 1. The coefficients that occur for **k = 1** are just the same as the first row of Pascal's triangle: namely, 1 and 1. The coefficients that occur for **k = 2** are just the same as the second row of Pascal's triangle: namely 1, 2, 1. And so forth. Of course if we think about how the binomial expression $(a + b)^k$ is multiplied out, then we see that the coefficients are formed by the very same rule that forms Pascal's triangle. And that explains why the rows of Pascal's triangle give the binomial coefficients.

Another remarkable fact is that the sum of the numbers in the kth row of Pascal's triange is 2^k. For example, in the third row, $1 + 3 + 3 + 1 = 8 = 2^3$. This is again a fundamental property of the binomial coefficients that can be verified with mathematical induction.

A pleasing interpretation of the rows of Pascal's triangle can be given in terms of coin tosses:

- If we toss a coin once, then there are two possible outcomes: one heads and one tails. This is information tabulated in the first row one of Pascal's triangle.

- If we toss a coin twice, then there are three possible outcomes: two heads (which can occur just one way), a head and a tail (which can occur two ways—heads-tails or tails-heads), and two tails (which can occur just one way). This information is tabulated by $1 - -2 - -1$ in the second row of Pascal's triangle.

- If we toss a coin three times, then there are four possible outcomes: three heads (which can occur just one way), two heads and a tail (which can occur three ways—heads-heads-tails, heads-tails-heads, or tails-heads-heads), two tails and a head (which can occur three ways—tails-tails-heads, tails-heads-tails, or heads-tails-tails), and three tails (which can occur just one way). This information is tabulated by $1 - -3 - -3 - -1$ in the third row of Pascal's triangle.

- And so forth.

Pascal's triangle is also a useful mnemonic for carrying information about the choose function. For this purpose we number the rows 0, 1, 2, and so on as usual. We also number the *terms* in each row 0 ,1, 2, and so on. Now suppose we wish to know how many different ways we can choose three objects from among five (in fact this came up in an example in the last section—the answer was 10). Simply go to row five of the triangle, term three, and we see the answer to be 10. Or if we want to know how many different ways to choose five items from among five (the answer is obviously 1), we go to row five and look at the fifth term (remembering to count from 0).

Of course Pascal's triangle is not magic—it is mathematics. And the mathematical explanation behind everything we have said here is the fundamental formula

$$\binom{n}{k} = \binom{n-1}{k-1} + \binom{n-1}{k}$$

You may test this formula by hand, or on your calculator. It is easy to confirm rigorously using mathematical induction. And it simply says (if we think of the kth element in the nth row of Pascal's triangle as a_{nk}) that $a_{nk} = a_{(n-1)(k-1)} + a_{(n-1)k}$. This just says that the element a_{nk} is formed by adding the two elements in the row above it.

6.9 Ramsey Theory

Frank Plumpton Ramsey (1903–1930) was a Professor at Cambridge University in England. In his tragically short life he established himself as an important and influential mathematician. He studied Whitehead and Russell's *Principia Mathematica* and offered a number of improvements, including ways to address Russell's paradox. He wrote just one paper, entitled "On a Problem of Formal Logic" (written in the year of his death) on the topic discussed here. The topic that we now call *Ramsey theory* is today a keystone of combinatorial theory, and is important for many parts of mathematics.

The general idea of Ramsey theory is to endeavor to find order in a set that is highly disordered. We will first illustrate the key idea with a popular example, and then we can discuss some of the more general principles. Let us consider an ordered sequence of questions:

- How many people need to be present at a party in order to guarantee that there will be two people who are acquainted or two people who are not? This is a very simple question, and the answer is two. If there are two people at the party, then they either know each other or they do not. End of discussion.
- How many people need to be present at a party in order to guarantee that there will be three people who are acquainted or two people who are not? This question is not quite so obvious. In fact five people will *not* do the trick (nor will four, three, or two people). To see this, it is useful and instructive to translate the question into one of graph theory.

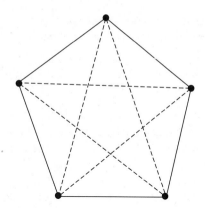

Figure 6.11   A graph on five vertices with no solid triangle and no dashed triangle.

Imagine that each person is represented by a point in the plane (which we think of as a vertex of a graph—see Chap. 8). Connect two points by a solid line if those two people are acquainted. Connect two points by a dashed line if those two people are not acquainted. Note that any two points will be connected by either a solid line or a dotted line, because every pair of people is either acquainted or not. So the question is: Can we always be sure that, no matter how the solid lines and the dashed lines are configured, there will always be a solid triangle or a dashed triangle? (Note here that a solid triangle corresponds to three people who are mutually acquainted. A dashed triangle corresponds to three people who are mutually unacquainted.) As we have said, for five vertices the answer is no. This fact is illustrated in Fig. 6.11.

It turns out that six is the magic number in order to guarantee three mutual acquaintances or three mutual unacquaintances. (Or, in other words, in a complete graph on six vertices consisting of solid lines and dashed lines, there will always be either a solid triangle or a dashed triangle.) To see this, imagine that one person in the party of six is named Astrid. There are five other people besides Astrid. Either she is acquainted with three of those or she is unacquainted with three of those. If she is acquainted with three, then examine those three. Either two of them know each other, or not. If two of them know each other, then those two plus Astrid form a mutually acquainted trio, just as we seek. If none of the three knows each other then that is a mutually unacquainted trio, just as we seek. If instead Astrid is unacquainted with three members of the party, then examine those three. If two of them are unacquainted, then Astrid and those two form a mutually unacquainted trio. If instead the three are all acquainted, then those three form a mutually acquainted trio. We see that, no matter what, there will be either a mutually acquainted trio or a mutually unacquainted trio.

- Now how many people will it take at a party for there to be a mutually acquainted quartet or a mutually unacquainted quartet? This is a *much* harder problem. It can be shown that the right number is 18.

- In order to form a party so that there will be either five mutually acquainted people or five mutually unacquainted people it is known that the right size for the party is between 43 and 49 inclusive. But the correct value is not known. This problem is currently considered to be beyond our computing power.

We list in a table now the known information about how many people r will be needed at a party in order to guarantee either k acquaintances or k unacquaintances. Note that, in our discussion above, we determined that when $k = 2$ then $r = 2$, when $k = 3$ then $r = 6$, when $k = 4$ then $r = 18$, and so forth. After $k = 4$, we indicate ranges of values for r because that is all that is known.

k	r
2	2
3	6
4	18
5	[43, 49]
6	[102, 165]
7	[205, 540]
8	[282, 1870]
9	[565, 6588]
10	[798, 23556]
11	[1597, 184755]
12	[1637, 705431]

Here is a simple interpretation of the third line of our table in terms of Internet sites. Suppose we look at 18 randomly chosen Websites or URLs. Then either four of them are about the same topic or four of them are about four mutually distinct topics. Another way to look at the matter is this: among 18 randomly chosen Websites, either four of them are all linked to each other, or four of them have no mutual links.

Exercises

1. There are 300 adult people in a room, none of them obese. Explain why two of them must have the same weight (in whole numbers of pounds).

2. There are 50 people in a room, none of them obese. Explain why two of them must have the same waist measurement (in whole numbers of inches).

3. Explain why the answer to Exercise 2 changes if the waist measurement is changed to whole numbers of millimeters.

4. There are 20 people sitting in a waiting room. The functionary in charge must choose five of these people to go to the green sanctuary and three of these people to go to the red sanctuary. In how many different ways can she do this?

5. In a standard deck of 52 playing cards, in how many different ways can you form two-of-a-kind?

6. In a standard deck of 52 playing cards, in how many different ways can you form four-of-a-kind?

7. A standard die used for gambling is a six-sided cube, with the sides numbered 1 through 6. You usually roll two dice at a time, and the two face-up values are added together to give your score. What is the likelihood that you will roll a seven?

8. Refer to the last exercise for terminology. What is the chance that you will roll two dice and get a two? How about a 12? Are there any other values that give this same answer? Why or why not?

9. Again refer to Exercise 7 for terminology. Now suppose that you are rolling three dice. Your score is obtained by adding together the three face values. What is your probability of getting a 10?

10. Solve the recursion $a_0 = 3$, $a_1 = -5$, and $a_j = a_{j-1} + 2a_{j-2}$ for $j \geq 2$.

11. Suppose we take a finite collection of points in the plane and connect every pair with an edge. We color each edge either red or blue or green. Will there be a triangle of just one color? With 16 points the answer is "no" but with 17 points the answer is "yes." Discuss.

CHAPTER 7

Matrices

7.1 What Is a Matrix?

A matrix is a rectangular array of numbers or variables or other algebraic objects.
An example of a matrix is

$$\begin{pmatrix} a & b & c & d & e \\ f & g & h & i & j \\ k & l & m & n & o \\ p & q & r & s & t \end{pmatrix}$$

We call this a 4×5 matrix because it has four rows and five columns. In general,
an $m \times n$ matrix has m rows and n columns.

When Arthur Cayley (1821–1895) invented matrices in the late nineteenth century, he bragged that he invented something that was of no earthly use. Rarely has
a person's assessment of his own work been more inaccurate. Today matrices are
used in all parts of mathematics, in engineering and physics, in the social sciences,
in statistics, and in any part of analytical thought where it is necessary to keep track

of (and to manipulate) information. What is important about matrices is that they can be *combined* in a number of useful ways—addition, multiplication, inversion, composition, and others—and each of these operations has significance for the information that the matrix contains. We shall learn a bit about these ideas in the present chapter.

7.2 Fundamental Operations on Matrices

We typically denote a matrix by a capital roman letter like A or M. The *elements* of the matrix A are designated by a_{ij}, where i is the *row* in which the element is located and j is the *column* in which the element is located. To take a specific example, consider the matrix

$$A = \begin{pmatrix} 3 & -1 & 4 & 2 \\ -6 & 5 & 4 & 0 \\ 1 & 9 & 14 & -8 \end{pmatrix}$$

For this matrix, $a_{23} = 4$ because the element of the matrix that is in the second row and third column is 4. Likewise, $a_{32} = 9$ and $a_{33} = 14$. Notice that, for this particular matrix, there are elements a_{ij} for $1 \leq i \leq 3$ (because there are three rows) and for $1 \leq j \leq 4$ (because there are four columns).

We can add two matrices only when they have the same size. If $A = (a_{ij})$ is an $m \times n$ matrix and $B = (b_{ij})$ is another $m \times n$ matrix, then their sum is $A + B = (a_{ij} + b_{ij})$ or

$$A + B = (a_{ij}) + (b_{ij}) = (a_{ij} + b_{ij})$$

EXAMPLE 7.1

As a concrete illustration of matrix addition, let

$$A = \begin{pmatrix} 3 & -6 \\ -2 & 4 \\ 1 & 0 \end{pmatrix} \quad \text{and} \quad B = \begin{pmatrix} -5 & 3 \\ 1 & -6 \\ 0 & 11 \end{pmatrix}$$

Then

$$A + B = \begin{pmatrix} 3 & -6 \\ -2 & 4 \\ 1 & 0 \end{pmatrix} + \begin{pmatrix} -5 & 3 \\ 1 & -6 \\ 0 & 11 \end{pmatrix} = \begin{pmatrix} -2 & -3 \\ -1 & -2 \\ 1 & 11 \end{pmatrix} \qquad \square$$

EXAMPLE 7.2
Matrices of the same size can also be subtracted. With the same A and B as in the previous example, we have

$$A - B = \begin{pmatrix} 3 & -6 \\ -2 & 4 \\ 1 & 0 \end{pmatrix} - \begin{pmatrix} -5 & 3 \\ 1 & -6 \\ 0 & 11 \end{pmatrix} = \begin{pmatrix} 8 & -9 \\ -3 & 10 \\ 1 & -11 \end{pmatrix} \qquad \square$$

Next we turn to multiplication of matrices. This is a bit more subtle than addition, and the form it takes may be something of a surprise. A first guess might be that the product of matrices $A = (a_{ij})$ and $B = (b_{ij})$ will be $A \cdot B = (a_{ij} \cdot b_{ij})$. In other words, we guess that we would multiply matrices componentwise. This turns out *not* to be a satisfactory way to define matrix multiplication, for it would result in equations like

$$\begin{pmatrix} 0 & 1 \\ 1 & 0 \end{pmatrix} \cdot \begin{pmatrix} 1 & 0 \\ 0 & 1 \end{pmatrix} = \begin{pmatrix} 0 & 0 \\ 0 & 0 \end{pmatrix}$$

In other words, the product of two nonzero matrices would be zero. This is not an attractive turn of events. We need a definition of matrix multiplication that avoids such problems, and also one that will preserve the essential information that is carried by the component matrices. These considerations motivate the definition that we are about to present.

The key fact about matrix multiplication is that we can multiply a matrix A times a matrix B (*in that order*) provide that A is an $m \times n$ matrix and B is an $n \times k$ matrix. In other words, the number of columns in A must match the number of rows in B. As an instance, if

$$A = \begin{pmatrix} 2 & -1 \\ -5 & 4 \\ 6 & 2 \end{pmatrix} \qquad \text{and} \qquad B = \begin{pmatrix} -5 & 2 & 6 & 9 \\ 4 & -3 & 8 & 12 \end{pmatrix}$$

then we may calculate $A \cdot B$ (because the number of columns in A matches the number of rows in B) but we may *not* calculate $B \cdot A$ (because the number of columns in B is four while the number of rows in A is three and these do *not* match).

Now how do we *actually calculate* a matrix product? If $A = (a_{ij})_{\substack{1 \le i \le m \\ 1 \le j \le n}}$ and $B = (b_{rs})_{\substack{1 \le r \le n \\ 1 \le s \le p}}$ (so that the number of columns in A matches the number of rows in B)

then we set $C = (c_{tu}) = A \cdot B$ and we define

$$c_{tu} = \sum_{k=1}^{n} a_{tk} b_{ku}$$

This is a all a bit abstract, so let us look at a concrete example.

EXAMPLE 7.3
Let

$$A = \begin{pmatrix} -3 & 2 & 10 \\ 6 & -4 & 9 \end{pmatrix} \quad \text{and} \quad B = \begin{pmatrix} 1 & 1 & -3 & 6 \\ 2 & -2 & 1 & 4 \\ 6 & -5 & -2 & 9 \end{pmatrix}$$

Let $C = (c_{tu}) = A \cdot B$. According to the rule,

$$c_{11} = \sum_{k=1}^{3} a_{1k} b_{k1} = (-3) \cdot 1 + 2 \cdot 2 + 10 \cdot 6 = 61$$

Likewise

$$c_{12} = \sum_{k=1}^{3} a_{1k} b_{k2} = (-3) \cdot 1 + 2 \cdot (-2) + 10 \cdot (-5) = -57$$

We calculate the other entries of $C = A \cdot B$ in a similar fashion. The final answer is

$$C = A \cdot B = \begin{pmatrix} 61 & -57 & -9 & 80 \\ 52 & -31 & -40 & 101 \end{pmatrix}$$

Notice here that A is a 2×3 matrix and B is a 3×4 matrix. Thus we mentally cancel the matching 3s and see that the product must be a 2×4 matrix. □

EXAMPLE 7.4
Let

$$A = \begin{pmatrix} 1 & 2 & 3 \\ 6 & 5 & 4 \\ -2 & 0 & -8 \end{pmatrix}$$

and

$$B = \begin{pmatrix} 3 & 2 & 1 \\ 4 & 5 & 6 \\ -8 & 0 & -2 \end{pmatrix}$$

Calculate both $A \cdot B$ and $B \cdot A$.

Solution: Now

$$A \cdot B = \begin{pmatrix} -13 & 12 & 7 \\ 6 & 37 & 28 \\ 58 & -4 & 14 \end{pmatrix}$$

Also

$$B \cdot A = \begin{pmatrix} 13 & 16 & 9 \\ 22 & 33 & -16 \\ -4 & -16 & -8 \end{pmatrix}$$

One immediate lesson here is that $A \cdot B \neq B \cdot A$. Multiplication of matrices is *not* commutative. A second lesson is that the only time we can calculate both $A \cdot B$ and $B \cdot A$ is when both matrics are square and both are of the same size.   □

7.3 Gaussian Elimination

In high school algebra everyone learns how to solve a system of two linear equations in two unknowns:

$$ax + by = \alpha$$

$$cx + dy = \beta$$

You simply multiply the first equation by a constant so that its x-coefficient matches the x-coefficient of the second equation. Then subtraction eliminates the x-variable and one can solve directly for y. Reverse substitution then yields the value of the x-variable, and the system is solved.

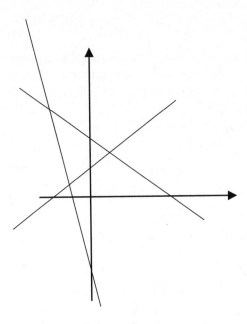

Figure 7.1 Typical (empty) intersection of three lines in the plane.

Matters are more complicated for systems of three equations in three unknowns, or more generally k equations in k unknowns. It becomes difficult to keep track of the information, and the attendant calculations can be daunting. In this section we introduce the method of *Gaussian elimination*, which gives a straightforward and virtually mechanical method for solving systems of linear equations. In fact it is straightforward to implement the method of *Gaussian elimination* on a computer; for a system of k equations in k unknowns it takes about k^3 calculations to find the solution. So this is a robust and efficient technique.

We concentrate our efforts on systems of k equations in k unknowns because such a system will generically have a unique solution (that is, a single value for each of the variables that solves the system). When there are *fewer* unknowns than equations then the system will typically have no solutions.[1] When there are *more* unknowns than equations then the system will typically have an entire *space* of solutions. For example, the solution of the system

$$x + z = 4$$

$$z = 3$$

---

[1]Think of intersecting lines in the plane. Two lines in the plane will usually have a point of intersection—as long as they are not parallel. Three lines in the plane will usually *not* have a mutual point of intersection—see Fig. 7.1.

are all points of the form $(1, y, 3)$. In other words, the set of solutions forms an entire line.

It requires ideas from linear algebra to handle these matters properly, and we cannot treat them here. So we will focus our attention on k equations in k unknowns.

In the method of *Gaussian elimination* we will typically have a system of equations

$$a_1^1 x_1 + a_2^1 x_2 + a_3^1 x_3 + \cdots + a_{k^1} x_k = \alpha^2$$

$$a_1^2 x_1 + a_2^2 x_2 + a_3^2 x_3 + \cdots + a_{k^2} x_k = \alpha^2$$

$$\cdots$$

$$a_1^k x_1 + a_2^k x_2 + a_3^k x_3 + \cdots + a_{k^k} x_k = \alpha^k$$

We will study this system by creating the associated *augmented matrix*

$$\begin{bmatrix} a_1^1 & a_2^1 & a_3^1 & \cdots & a_k^1 & \alpha^1 \\ a_1^2 & a_2^2 & a_3^2 & \cdots & a_k^2 & \alpha^2 \\ & & \cdots & & & \\ a_1^k & a_2^k & a_3^k & \cdots & a_k^k & \alpha^k \end{bmatrix}$$

Now there are certain allowable operations on the augmented matrix that we use to reduce it to a normalized form. The *main diagonal* of this matrix is the line of terms given by $a_1^1, a_2^2, \ldots, a_k^k$. We want all the terms *below* this main diagonal to be equal to 0. And we want the terms on the main diagonal to all be 1s. The three allowable operations are:

1. Switch the position of two rows.
2. Multiply any row by a nonzero constant.
3. Add a multiple of one row to another.

It turns out that these three simple moves are always adequate to do the job. The ideas are best illustrated with some examples.

EXAMPLE 7.5
Let us solve the system

$$x - 2y + 3z = 4$$

$$x + y + z = -2$$

$$-x + 2y + z = 1$$

using Gaussian elimination. We begin by writing the augmented matrix:

$$\left[\begin{array}{ccc|c} 1 & -2 & 3 & 4 \\ 1 & 1 & 1 & -2 \\ -1 & 2 & 1 & 1 \end{array}\right]$$

Keeping in mind that our aim is to produce all zeros below the main diagonal, we subtract the first row from the second. This yields

$$\left[\begin{array}{ccc|c} 1 & -2 & 3 & 4 \\ 0 & 3 & -2 & -6 \\ -1 & 2 & 1 & 1 \end{array}\right]$$

This produces a zero below the main diagonal.

Next we add the first row to the last. The result is

$$\left[\begin{array}{ccc|c} 1 & -2 & 3 & 4 \\ 0 & 3 & -2 & -6 \\ 0 & 0 & 4 & 5 \end{array}\right]$$

Finally let us multiply the second row by $1/3$ and the third row by $1/4$. The end result is

$$\left[\begin{array}{ccc|c} 1 & -2 & 3 & 4 \\ 0 & 1 & -2/3 & -2 \\ 0 & 0 & 1 & 5/4 \end{array}\right]$$

This is the normalized form that we seek.

Writing our information again as a linear system, we have

$$x - 2y + 3z = 4$$

$$0x + 1y - 2/3z = -2$$

$$0x + 0y + z = 5/4$$

We may immediately read off from the last equation that $z = 5/4$. Substituting this information into the second equation gives $y = -7/6$. Lastly, putting both these values into the first equation yields $x = -25/12$. Thus the solution of our system is $x = -25/12$, $y = -7/6$, $z = 5/4$ or $(-25/12, -7/6, 5/4)$.

We encourage you to check this solution by plugging the values into the three equations of the original system.   □

This was so easy that we are granted the courage now to attack a system of four equations in four unknowns. It is gratifying how straightforward the procedure is:

EXAMPLE 7.6
Let us use Gaussian elimination to solve the system of four equations in four unknowns given by

$$2x - y + 3z + w = 2$$

$$x + y - z - w = -1$$

$$x + z + w = 4$$

$$-x - y + 2w = 1$$

As before, our first step is to write the associated augmented matrix:

$$\left[\begin{array}{rrrr|r} 2 & -1 & 3 & 1 & 2 \\ 1 & 1 & -1 & -1 & -1 \\ 1 & 0 & 1 & 1 & 4 \\ -1 & -1 & 0 & 2 & 1 \end{array}\right]$$

Notice that we were careful in the third row to put a 0 in the second position because the third equation has no y. Likewise we put a 0 in the third position of the fourth row because the fourth equation has no z.

Let us begin by multiplying the first row by $1/2$. Thus we have

$$\left[\begin{array}{rrrr|r} 1 & -1/2 & 3/2 & 1/2 & 1 \\ 1 & 1 & -1 & -1 & -1 \\ 1 & 0 & 1 & 1 & 4 \\ -1 & -1 & 0 & 2 & 1 \end{array}\right]$$

Now we substract the first row from the second row and the third row (we are performing two operations at once). Thus

$$\left[\begin{array}{rrrr|r} 1 & -1/2 & 3/2 & 1/2 & 1 \\ 0 & 3/2 & -5/2 & -3/2 & -2 \\ 0 & 1/2 & -1/2 & 1/2 & 3 \\ -1 & -1 & 0 & 2 & 1 \end{array}\right]$$

Next we add the first row to the last. So

$$\begin{bmatrix} 1 & -1/2 & 3/2 & 1/2 & | & 1 \\ 0 & 3/2 & -5/2 & -3/2 & | & -2 \\ 0 & 1/2 & -1/2 & 1/2 & | & 3 \\ 0 & -3/2 & 3/2 & 5/2 & | & 2 \end{bmatrix}$$

We see now the advantage of working systematically (that is, normalizing the first column first). For now we can add the second row to the fourth row. Therefore

$$\begin{bmatrix} 1 & -1/2 & 3/2 & 1/2 & | & 1 \\ 0 & 3/2 & -5/2 & -3/2 & | & -2 \\ 0 & 1/2 & -1/2 & 1/2 & | & 3 \\ 0 & 0 & -1 & 1 & | & 0 \end{bmatrix}$$

Now we subtract 1/3 the second row from the third. The result is

$$\begin{bmatrix} 1 & -1/2 & 3/2 & 1/2 & | & 1 \\ 0 & 3/2 & -5/2 & -3/2 & | & -2 \\ 0 & 0 & 1/3 & 1 & | & 11/3 \\ 0 & 0 & -1 & 1 & | & 0 \end{bmatrix}$$

The last step is to add three times the third row to the fourth. Hence

$$\begin{bmatrix} 1 & -1/2 & 3/2 & 1/2 & | & 1 \\ 0 & 3/2 & -5/2 & -3/2 & | & -2 \\ 0 & 0 & 1/3 & 1 & | & 11/3 \\ 0 & 0 & 0 & 4 & | & 11 \end{bmatrix}$$

We conclude by multiplying each row by a suitable constant so that the lead term is 1. We finally have the augmented matrix

$$\begin{bmatrix} 1 & -1/2 & 3/2 & 1/2 & | & 1 \\ 0 & 1 & -5/3 & -1 & | & -4/3 \\ 0 & 0 & 1 & 3 & | & 11 \\ 0 & 0 & 0 & 1 & | & 11/4 \end{bmatrix}$$

Translating back to a linear system, we have

$$1x - 1/2y + 3/2z + 1/2w = 1$$

$$0x + 1y - 5/3z - 1w = -4/3$$

$$0x + 0y + 1z + 3w = 11$$

$$0x + 0y + 0z + w = 11/4$$

We see immediately that $w = 11/4$. Back substituting, we find that $z = 11/4$. Again back substituting, we find that $y = 6$. Finally, the first equation tells us that $x = -3/2$. So the solution set is $x = -3/2$, $y = 6$, $z = 11/4$, $w = 11/4$ or $(-3/2, 6, 11/4, 11/4)$.   □

7.4 The Inverse of a Matrix

The matrix

$$I = \begin{pmatrix} 1 & 0 & 0 & 0 & \cdots & 0 & 0 \\ 0 & 1 & 0 & 0 & \cdots & 0 & 0 \\ 0 & 0 & 1 & 0 & \cdots & 0 & 0 \\ & & & \cdots & & & \\ 0 & 0 & 0 & 0 & \cdots & 1 & 0 \\ 0 & 0 & 0 & 0 & \cdots & 0 & 1 \end{pmatrix}$$

is called the *identity matrix*. It has this name because if we multiply it with any square matrix A of the same size we obtain $I \cdot A = A \cdot I = A$. For example, if

$$A = \begin{pmatrix} a & b & c \\ d & e & f \\ g & h & i \end{pmatrix}$$

then

$$I \cdot A = \begin{pmatrix} 1 & 0 & 0 \\ 0 & 1 & 0 \\ 0 & 0 & 1 \end{pmatrix} \cdot \begin{pmatrix} a & b & c \\ d & e & f \\ g & h & i \end{pmatrix} = \begin{pmatrix} a & b & c \\ d & e & f \\ g & h & i \end{pmatrix}$$

Likewise,

$$A \cdot I = \begin{pmatrix} a & b & c \\ d & e & f \\ g & h & i \end{pmatrix} \cdot \begin{pmatrix} 1 & 0 & 0 \\ 0 & 1 & 0 \\ 0 & 0 & 1 \end{pmatrix} = \begin{pmatrix} a & b & c \\ d & e & f \\ g & h & i \end{pmatrix}$$

So I plays the role of the multiplicative identity.

Let A be a given square matrix. So A has size $n \times n$ for some positive integer n. We seek a matrix A' so that $A \cdot A' = I$ and $A' \cdot A = I$. The matrix A' will play the role of the *multiplicative inverse* of A. We call it the *inverse matrix* to A.

Now it is a fact that not every matrix has a multiplicative inverse. Consider the example of

$$A = \begin{pmatrix} 1 & 0 \\ 0 & 0 \end{pmatrix}$$

If A' is any matrix,

$$A' = \begin{pmatrix} a & b \\ c & d \end{pmatrix}$$

then

$$A' \cdot A = \begin{pmatrix} a & 0 \\ c & 0 \end{pmatrix}$$

Thus $A' \cdot A$ cannot possibly be the identity matrix. A similar calculation shows that $A \cdot A'$ cannot possibly be the identity matrix. So the matrix A definitely does not have an inverse.

It is useful to have a simple and direct calculation that will tell us when a given matrix has an inverse and when it does not. And in fact there is such a calculation which we shall now learn. It is based on the Gaussian elimination technique that was presented in the last section.

The idea is best understood in the context of 2×2 matrices. Such a matrix which does *not* have an inverse is one in which the second row is a multiple of the first:

$$A = \begin{pmatrix} a & b \\ \lambda a & \lambda b \end{pmatrix}$$

If we guess that some matrix

$$A' = \begin{pmatrix} a' & b' \\ c' & d' \end{pmatrix}$$

is an inverse for A, then we would have

$$I = A \cdot A' = \begin{pmatrix} aa' + bc' & ab' + bd' \\ \lambda aa' + \lambda bc' & \lambda ab' + \lambda bd' \end{pmatrix}$$

We see in the product that the second row is a constant multiple λ times the first row. So the product matrix certainly *cannot* be the identity I, and we see that A cannot be invertible.

Thus, in the context of 2×2 matrices, the situation to rule out is that one row be a multiple of the other. It turns out that a convenient way to do this is by way of the determinant. If

$$A = \begin{pmatrix} a & b \\ c & d \end{pmatrix}$$

is a given matrix then its *determinant* is given by

$$\det A = ad - bc$$

The two rows are *not* multiples of each other precisely when $\det A \neq 0$. This will be precisely the circumstance in which we will be able to find an inverse for the matrix A.

EXAMPLE 7.7
Let

$$M = \begin{pmatrix} -5 & 4 \\ 2 & -6 \end{pmatrix}$$

Calculate $\det M$.   □

Solution: We see that

$$\det M = (-5) \cdot (-6) - 2 \cdot 4 = 22 \neq 0$$ □

The situation for a 3×3 matrix is analogous. Let

$$A = \begin{pmatrix} a & b & c \\ d & e & f \\ g & h & i \end{pmatrix}$$

be a 3×3 matrix. The determinant is defined to be

$$\det A = a \cdot \det \begin{pmatrix} e & f \\ h & i \end{pmatrix} - b \cdot \det \begin{pmatrix} d & f \\ g & i \end{pmatrix} + c \cdot \det \begin{pmatrix} d & e \\ g & h \end{pmatrix}$$

This is a natural sort of generalization of the determinant for 2×2 matrices. We call

$$\begin{pmatrix} e & f \\ h & i \end{pmatrix}$$

the *minor* associated to the entry a. Likewise we call

$$\begin{pmatrix} d & f \\ g & i \end{pmatrix}$$

the minor associated to the entry b and

$$\begin{pmatrix} d & e \\ g & h \end{pmatrix}$$

the minor associated to the entry c.

What we have done here is to expand the determinant *by the first row*. It is also possible to expand the determinant by any row or column, just so long as we assign the right plus or minus signs to each of the components. For the purposes of the present book, expansion by the first row will suffice.

EXAMPLE 7.8
Let

$$B = \begin{pmatrix} -7 & 4 & -3 \\ 2 & -5 & 3 \\ 1 & -3 & -2 \end{pmatrix}$$

Calculate $\det B$. □

Solution: We calculate that

$$\det B = (-7) \cdot \det \begin{pmatrix} -5 & 3 \\ -3 & -2 \end{pmatrix}$$

$$- 4 \cdot \det \begin{pmatrix} 2 & 3 \\ 1 & -2 \end{pmatrix}$$

$$+ (-3) \cdot \det \begin{pmatrix} 2 & -5 \\ 1 & -3 \end{pmatrix}$$

$$= (-7) \cdot 19 - 4 \cdot (-7) + (-3) \cdot (-1)$$

$$= -102$$

$$\neq 0. \qquad\qquad \square$$

It is worth summarizing the main rule for inverses for which we have given an indication of the justification:

> **Rule for inverses:** A square matrix has an inverse if and only if the matrix has nonzero determinant.

This still does not tell us how to *find* the inverse, but we shall get to that momentarily.

There are in fact a number of ways to calculate the inverse of a matrix, and we shall indicate two of them here. The first is the method of Gaussian elimination, which is a powerful technique that can be used for many purposes. The idea is to take the given square matrix

$$A = \begin{pmatrix} a & b \\ c & d \end{pmatrix}$$

and to *augment* it by adjoining the identity matrix as in the display below:

$$\left(\begin{array}{cc|cc} a & b & 1 & 0 \\ c & d & 0 & 1 \end{array} \right)$$

Now the method of gaussian elimination allows us to perform certain operations on the rows of this augmented matrix, the goal being to reduce the square matrix on the left to the identity matrix. Whatever matrix results on the right will be the inverse.

The three allowable operations are:

(1) We can switch the position of any two rows.

(2) We can multiply any given row by a constant.

(3) We can add any multiple of one row to another row.

Let us illustrate the method with a concrete example.

EXAMPLE 7.9
Let

$$M = \begin{pmatrix} 2 & 4 \\ -3 & 5 \end{pmatrix}$$

Then $\det M = 2 \cdot 5 - (-3) \cdot 4 = 22 \neq 0$. Hence this matrix passes the test, and it has an inverse.

We now examine the augmented matrix

$$\left(\begin{array}{cc|cc} 2 & 4 & 1 & 0 \\ -3 & 5 & 0 & 1 \end{array} \right)$$

Our job is to perform Gaussian elimination and to thereby transform the left-hand square matrix into the identity matrix.

We begin by multiplying the first row through by $1/2$. The result is

$$\left(\begin{array}{cc|cc} 1 & 2 & 1/2 & 0 \\ -3 & 5 & 0 & 1 \end{array} \right)$$

Now we add three times the first row to the second. We obtain the augmented matrix

$$\left(\begin{array}{cc|cc} 1 & 2 & 1/2 & 0 \\ 0 & 11 & 3/2 & 1 \end{array} \right)$$

What we have achieved thus far is that the first column of the left-hand square matrix looks like the first column of the identity.

Now we subtract $2/11$ of the second row from the first. The result is

$$\left(\begin{array}{cc|cc} 1 & 0 & 5/22 & -2/11 \\ 0 & 11 & 3/2 & 1 \end{array} \right)$$

Finally we multiply the second row by $1/11$. In the end, then, we obtain

$$\begin{pmatrix} 1 & 0 & 5/22 & -2/11 \\ 0 & 1 & 3/22 & 1/11 \end{pmatrix}$$

What we read off from this last augmented matrix is that the inverse of the original matrix M is

$$M^{-1} = \begin{pmatrix} 5/22 & -2/11 \\ 3/22 & 1/22 \end{pmatrix}$$

We invite you to test the result by multiplying $M \times M^{-1}$.   □

EXAMPLE 7.10
Let us calculate the inverse of the matrix

$$C = \begin{pmatrix} 3 & 0 & 1 \\ 0 & 1 & 2 \\ 1 & 0 & 1 \end{pmatrix}$$

We begin by calculating the determinant to make sure that the matrix C is invertible:

$$\det C = 3 \det \begin{pmatrix} 1 & 2 \\ 0 & 1 \end{pmatrix} - 0 \det \begin{pmatrix} 0 & 2 \\ 1 & 1 \end{pmatrix} + 1 \det \begin{pmatrix} 0 & 1 \\ 1 & 0 \end{pmatrix}$$

$$= 3 \cdot 1 - 0 \cdot (-2) + 1 \cdot (-1) = 2 \neq 0$$

Thus C passes the test and we may compute the inverse matrix.
    We write the augmented matrix as

$$\begin{pmatrix} 3 & 0 & 1 & 1 & 0 & 0 \\ 0 & 1 & 2 & 0 & 1 & 0 \\ 1 & 0 & 1 & 0 & 0 & 1 \end{pmatrix}$$

As a first step, we multiply the first row by $1/3$:

$$\begin{pmatrix} 1 & 0 & 1/3 & 1/3 & 0 & 0 \\ 0 & 1 & 2 & 0 & 1 & 0 \\ 1 & 0 & 1 & 0 & 0 & 1 \end{pmatrix}$$

Next we subtract the first row from the third row:

$$\begin{pmatrix} 1 & 0 & 1/3 & \bigm| & 1/3 & 0 & 0 \\ 0 & 1 & 2 & \bigm| & 0 & 1 & 0 \\ 0 & 0 & 2/3 & \bigm| & -1/3 & 0 & 1 \end{pmatrix}$$

Now we subtract three times the third row from the second row:

$$\begin{pmatrix} 1 & 0 & 1/3 & \bigm| & 1/3 & 0 & 0 \\ 0 & 1 & 0 & \bigm| & 1 & 1 & -3 \\ 0 & 0 & 2/3 & \bigm| & -1/3 & 0 & 1 \end{pmatrix}$$

As a penultimate step, we multiply the last row by $3/2$. The result is:

$$\begin{pmatrix} 1 & 0 & 1/3 & \bigm| & 1/3 & 0 & 0 \\ 0 & 1 & 0 & \bigm| & 1 & 1 & -3 \\ 0 & 0 & 1 & \bigm| & -1/2 & 0 & 3/2 \end{pmatrix}$$

Finally, we subtract one-third the last row from the first row. We obtain:

$$\begin{pmatrix} 1 & 0 & 0 & \bigm| & 1/2 & 0 & -1/2 \\ 0 & 1 & 0 & \bigm| & 1 & 1 & -3 \\ 0 & 0 & 1 & \bigm| & -1/2 & 0 & 3/2 \end{pmatrix}$$

Thus the inverse matrix to the original matrix C is

$$C^{-1} = \begin{pmatrix} 1/2 & 0 & -1/2 \\ 1 & 1 & -3 \\ -1/2 & 0 & 3/2 \end{pmatrix}$$

As usual, we may check our work:

$$C \cdot C^{-1} = \begin{pmatrix} 3 & 0 & 1 \\ 0 & 1 & 2 \\ 1 & 0 & 1 \end{pmatrix} \cdot \begin{pmatrix} 1/2 & 0 & -1/2 \\ 1 & 1 & -3 \\ -1/2 & 0 & 3/2 \end{pmatrix} = \begin{pmatrix} 1 & 0 & 0 \\ 0 & 1 & 0 \\ 0 & 0 & 1 \end{pmatrix}$$

□

7.5 Markov Chains

A *Markov chain*—named after Andrei Andreyevich Markov (1856–1922)—is a random process in which future states of the process may depend on the present state but *not* on previous states. An example is the flip of a coin. You can flip a coin 100 times and get heads every time (it's not very likely, but it could happen). This tells you nothing about what the next flip will be: the chances are 50% heads and 50% tails and that is all there is to it. Many gambling fallacies and "systems" are based on a misunderstanding of this simple principle.

Another example of a Markov chain is the weather. The weather tomorrow may depend in part on the weather today, but it does not depend on the weather in the past.

Let us begin to understand how to analyze a Markov chain by examining a concrete example.

EXAMPLE 7.11

Suppose that we live in a climate in which a sunny day is 80% likely to be followed by another sunny day. And a rainy day is 40% likely to be followed by another rainy day. We represent the state of the weather on any given day by a vector (s, r) where s stands for the likelihood of sun and r stands for the likelihood of rain.

We may represent the stated probabilities (which are garnered from 5 years' observation of the weather in this region) with the matrix

$$P = \begin{pmatrix} 0.8 & 0.4 \\ 0.2 & 0.6 \end{pmatrix}$$

What does this mean? Suppose that we begin our analysis on the first day—call it day 1—on which it is sunny. Thus the vector representing the status today is $\mathbf{x}^{(1)} = (1, 0)$. Because it is definitely sunny and it is not raining. Now we calculate the next state—the weather tomorrow, which will be represented by a vector $\mathbf{x}^{(2)}$—by applying the given probabilities to $\mathbf{x}^{(1)}$. This is done simply by multiplying the vector $\mathbf{x}^{(1)}$ by the matrix P. Thus

$$\mathbf{x}^{(2)} = P \cdot \mathbf{x}^{(1)} = \begin{pmatrix} 0.8 & 0.4 \\ 0.2 & 0.6 \end{pmatrix} \cdot \begin{pmatrix} 1 \\ 0 \end{pmatrix} = \begin{pmatrix} 0.8 \\ 0.2 \end{pmatrix}$$

Notice that when we are in text we write our vectors (horizontally) as (s, r), but when we calculate with matrices we write the vectors vertically. This is customary in mathematics.

We have learned that, on the second day, there is 0.8 probability (or 80% probability) that it will be sunny and 0.2 probability (or 20% probability) that it will rain.

It is worth noting that the columns of the matrix P are nonnegative numbers that sum to 1. And the entries of our vectors are nonnegative numbers that sum to 1. This is because they are probabilities (which always lie between 0 and 1), and they represent all possible outcomes. The sum of the probabilities of all possible outcomes will of course be 1.

In the same fashion we can determine the likely weather on day 3. We see that

$$\mathbf{x}^{(3)} = P \cdot \mathbf{x}^{(2)} = P \cdot \begin{pmatrix} 0.8 \\ 0.2 \end{pmatrix} = \begin{pmatrix} 0.8 & 0.4 \\ 0.2 & 0.6 \end{pmatrix} \cdot \begin{pmatrix} 0.8 \\ 0.2 \end{pmatrix} = \begin{pmatrix} 0.72 \\ 0.28 \end{pmatrix}$$

We find that, on the third day, there is 0.72 probability that it will be sunny and 0.28 probability that it will rain. Again, the two probabilities add up to 1. ☐

As time goes on, the weather predictions become less and less accurate. We all know from experience that we cannot predict the weather on May 15 by examining the weather on April 15. Yet that is what we are doing in the last example: we predict the weather on day 2 by doing a calculation based on the weather values on day 1; we predict the weather on day 3 by doing a calculation with the weather values on day 2 (which were in turn determined from the weather values on day 1); and so forth. It is natural to wonder what "steady state" the weather may be tending to as time marches on to infinity.

Formulated mathematically, we want to calculate the limit

$$\mathbf{q} = \lim_{j \to \infty} \mathbf{x}^{(j)} \tag{7.1}$$

This can be rewritten as

$$\mathbf{q} = \lim_{j \to \infty} P^{j-1} \mathbf{x}^{(1)}$$

where the power of P indicates that the matrix P is applied (or multiplied in) that many times. Applying P to both sides of the equation, we find that

$$P\mathbf{q} = P \left[\lim_{j \to \infty} P^{j-1} \mathbf{x}^{(1)} \right] = \lim_{j \to \infty} P \cdot P^{j-1} \mathbf{x}^{(1)} = \lim_{j \to \infty} P^{j} \mathbf{x}^{(1)} = \mathbf{q}|$$

In the last equality we use the fact—recorded in (7.1)—that $\mathbf{q}$ is the limit of iterates of P applied to $\mathbf{x}^{(1)}$.

Thus we discover that the limiting, or steady state vector for our weather system is a vector $\mathbf{q}$ that is fixed by the matrix P. The operative equation is

$$P\mathbf{q} = \mathbf{q}$$

EXAMPLE 7.12
Let us find the steady state vector $\mathbf{q}$ for the weather system described in the last example.

We seek a vector $\mathbf{q} = (q_1, q_2)$ such that the q_j's are nonnegative and sum to 1, and so that

$$P\mathbf{q} = \mathbf{q}$$

This may be written out as

$$\begin{pmatrix} 0.8 & 0.4 \\ 0.2 & 0.6 \end{pmatrix} \cdot \begin{pmatrix} q_1 \\ q_2 \end{pmatrix} = \begin{pmatrix} q_1 \\ q_2 \end{pmatrix}$$

This translates to

$$\begin{pmatrix} 0.8q_1 + 0.4q_2 \\ 0.2q_1 + 0.6q_2 \end{pmatrix} = \begin{pmatrix} q_1 \\ q_2 \end{pmatrix}$$

or

$$0.8q_1 + 0.4q_2 = q_1$$

$$0.2q_1 + 0.6q_2 = q_2$$

Notice that these simplify to

$$-0.2q_1 + 0.4q_2 = 0$$

$$0.2q_1 - 0.4q_2 = 0$$

These two equations are redundant, as they are multiples of each other.
Thus we are reduced to solving

$$-0.2q_1 + 0.4q_2 = 0$$

alongside the condition

$$q_1 + q_2 = 1$$

This is easy to do, and the solution is $q_1 = 2/3 \approx 0.6666$ and $q_2 = 1/3 \approx 0.3333$.

We conclude that the steady state weather system is $(0.6666, 0.3333)$. In the long run, it is twice as likely that we will have sun as that it will rain. $\square$

You Try It: A rain forest has three types of trees: young (15–25 years), middle-aged (26–45 years), and old (46 years or older). Let $y(j)$, $m(j)$, and $o(j)$ denote the number of each type of tree, respectively. Let d_y, d_m, and d_o denote the loss rates (expressed in percent) of each type of tree. Let $b(j)$ denote the number of baby trees (less than 15 years old) in the forest and d_b the loss rate for baby trees.

Express $b(j + 1)$ in terms of $b(j)$, $y(j)$, $m(j)$, and $o(j)$ (and the percentages indicated above). Write down a matrix that represents that Markov process that is present in this system.

7.6 Linear Programming

Linear programming is a fundamental mathematical tool that has existed—at least in its modern form—since M. K. Wood and G. B. Dantzig's fundamental papers [WOD] and [DAN] in 1949. The basic idea is that one wants to reduce an impossibly complex—or computationally expensive—problem to a more tractable problem using some kind of analysis. Dantzig's method of linear programming is extraordinarily effective at this job. Today the major airlines use a version of the simplex method to schedule flights and many industries use linear programming to schedule job runs and other tasks. There are many standard computer packages that are dedicated to linear programming methods.

In the most basic linear programming setting, one is given an *objective function* f that is linear:

$$f(x_1, x_2, \ldots, x_N) = a_1 x_1 + a_2 x_2 + \cdots + a_N x_N$$

and one wishes to maximize (or minimize) this function as the variable point $(x_1, x_2, \ldots, x_N)$ varies over a region $\mathcal{R}$ in space that is defined by some linear inequalities (called *constraints*):

$$b_1^1 x_1 + b_2^1 x_2 + \cdots + b_N^1 x_N \leq \beta^1$$
$$b_1^2 x_1 + b_2^2 x_2 + \cdots + b_N^2 x_N \leq \beta^2$$
$$\cdots$$
$$b_1^k x_1 + b_2^k x_2 + \cdots + b_N^k x_N \leq \beta^k$$

Figure 7.2   A feasible region in space.

See Fig. 7.2. We call this region in space the *feasible region*. A linear inequality such as we see in the last display describes a halfspace. So the points that simultaneously satisfy all the linear inequalities will typically describe a convex polytope in space. That is what the figure suggests.

Now the graph of a linear function such as f above is just a plane or (in higher dimensions) a hyperplane. So picture such a graph lying over the feasible region $\mathcal{R}$. See Fig. 7.3. It is not difficult to picture that, in some directions, the graph of the linear function goes up hill. And there will be a particular direction in which

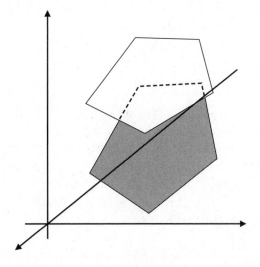

Figure 7.3   The graph of a linear function over a feasible region.

it goes uphill as steeply as possible. The function will reach its maximum value at that point on the graph which is as high as possible, and that will occur at a corner of the feasible region. This is the basic idea behind linear programming.

A detailed course in the theory of linear programming would teach you matrix techniques for finding the corner or vertex at which the maximum is achieved. This is actually quite an interesting process. In a typical application—such as scheduling flights for an airline—the feasible region will have thousands of vertices, and the objective function f will have hundreds or thousands of variables. So one certainly would not want to check every possibility by hand.

Of course a computer can help, but these problems are so large that even a computer gets bogged down with the calculations. The simplex method of Dantzig gives a very efficient way of zeroing in on the vertices where the extrema lie.

In this very brief treatment we cannot get into all the particulars of the linear programming method. But we can work some simple examples that illustrate the idea behind the solution of a linear programming problem.

EXAMPLE 7.13
Find the maximum value of the objective function

$$f(x, y) = 2x + 7y$$

over the feasible region

$$y - 2x \leq 3$$
$$y + 3x \geq -4$$
$$y - 10x \geq -2$$

$\square$

Solution: Figure 7.4 illustrates the feasible region.

The three points of intersection of the edges (that is, the vertices or corners) are easy to solve for using elementary techniques. These are $(-7/5, 1/5)$, $(5/8, 17/4)$, and $(-2/13, -46/13)$. Now we know from Dantzig's theory of linear programming (that is, the simplex method) that the extrema of the function f will occur at one of these three vertices. This problem is small enough so that it is tractable to just plug in the points and examine the values:

$$\left(-\frac{7}{5}\right), \left(\frac{1}{5}\right) = \frac{7}{5}$$

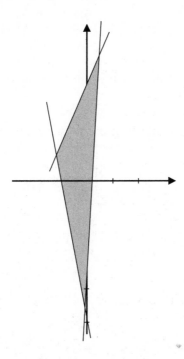

Figure 7.4   The feasible region for Example 7.13.

$$\left(\frac{5}{8}\right), \left(\frac{17}{4}\right) = 31$$

$$\left(-\frac{2}{13}\right), \left(\frac{46}{13}\right) = -\left(\frac{326}{13}\right)$$

Plainly the greatest value is 31, and it is assumed at the point $(5/8, 17/4)$. We may also note that the least value is $-326/13$, and it is assumed at the point $(-2/13, -46/13)$.   □

EXAMPLE 7.14

Find the extrema of the objective function

$$f(x, y, z) = 3x - 4y + 7z$$

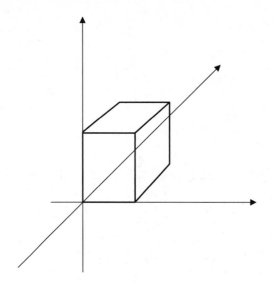

Figure 7.5   The feasible region for Example 7.14.

on the feasible region

$$x \leq 1$$

$$x \geq 0$$

$$y \leq 1$$

$$y \geq 0$$

$$z \leq 1$$

$$z \geq 0 \qquad \qquad \square$$

Solution: Figure 7.5 illustrates the feasible region. It is of course a cube.

It is no trouble at all to observe that the vertices (corners) are the points $(1, 0, 0)$, $(0, 1, 0)$, $(0, 0, 1)$, $(0, 1, 1)$, $(1, 0, 1)$, $(1, 1, 0)$, $(0, 0, 0)$, and $(1, 1, 1)$. We may calculate the values of f at those points directly:

$$f(1, 0, 0) = 3$$

$$f(0, 1, 0) = -4$$

$$f(0, 0, 1) = 7$$

$$f(0, 1, 1) = 3$$

$$f(1, 0, 1) = 10$$

$$f(1, 1, 0) = -1$$

$$f(0, 0, 0) = 0$$

$$f(1, 1, 1) = 6$$

We see that the greatest value of f is 10, and it is assumed at the point $(1, 0, 1)$. The least value of f is -4, and it is assumed at the point $(0, 1, 0)$. ☐

Exercises

In Exercises 1 and 2, say what the dimensions of the matrix are.

1. $\begin{pmatrix} 3 & 2 & 7 \\ 4 & -9 & -2 \\ 2 & 1 & 6 \\ 5 & 0 & 0 \end{pmatrix}$

2. $\begin{pmatrix} 3 & 5 & -4 & 2 & 0 \\ 1 & 1 & 9 & 4 & -2 \\ 6 & 4 & 3 & 2 & -1 \end{pmatrix}$

In Exercises 3, 4, 5, and 6, calculate the indicated matrix operation.

3. $\begin{pmatrix} 3 & -2 & 5 \\ 2 & 9 & 6 \end{pmatrix} + \begin{pmatrix} -6 & 2 & 8 \\ 7 & -4 & 9 \end{pmatrix}$

4. $\begin{pmatrix} 6 & -5 & 4 \\ 3 & 4 & 4 \end{pmatrix} \cdot \begin{pmatrix} 6 & -5 \\ 3 & 4 \\ -2 & 2 \end{pmatrix}$

5. $\begin{pmatrix} 5 & -2 & 3 \\ 1 & 4 & 9 \end{pmatrix} - \begin{pmatrix} 6 & 8 & -1 \\ 0 & 2 & 0 \end{pmatrix}$

6. $\begin{pmatrix} 2 & -3 \\ 4 & -7 \end{pmatrix} \cdot \begin{pmatrix} -5 & 1 \\ -9 & 6 \end{pmatrix}$

7. Calculate the inverse of the matrix

$$\begin{pmatrix} 1 & 0 & 1 \\ 0 & 1 & 1 \\ 2 & 1 & 0 \end{pmatrix}$$

8. Without actually calculating the inverse, say whether this matrix has an inverse:

$$\begin{pmatrix} 3 & 2 & 1 \\ -1 & -3 & -2 \\ 0 & 1 & 6 \end{pmatrix}$$

9. An unfair coin has twice the likelihood of landing heads as tails. Describe the coin flips as a Markov process.

CHAPTER 8

Graph Theory

8.1 Introduction

We learn even in high school about graphs of functions. The graph of a function is usually a curve drawn in the x-y plane. See Fig. 8.1. But the word "graph" has other meanings. In finite or discrete mathematics, a graph is a collection of points and edges or arcs in the plane. Fig. 8.2 illustrates a graph as we are now discussing the concept.

Leonhard Euler (1707–1783) is considered to have been the father of graph theory. His paper in 1736 on the seven bridges of Königsberg is considered to have been the foundational paper in the subject. It is worthwhile now to review that topic.

Königsberg is a town, founded in 1256, that was originally in Prussia. After a stormy history, the town became part of the Soviet Union and was renamed Kaliningrad in 1946. In any event, during Euler's time the town had seven bridges (named Krämer, Schmiede, Holz, Hohe, Honig, Köttel, and Grünespanning) spanning the Pregel River. Fig. 8.3 gives a simplified picture of how the bridges were originally configured (two of the bridges were later destroyed during World War II, and two others demolished by the Russians). The question that fascinated people

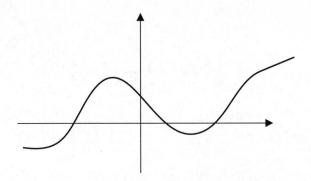

Figure 8.1　A graph of a function in the plane.

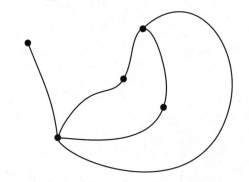

Figure 8.2　A graph as a combinatorial object.

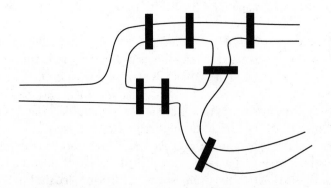

Figure 8.3　The seven bridges at Königsberg.

in the eighteenth century was whether it was possible to walk a route that never repeats any part of the path and that crosses each bridge exactly once.

Euler in effect invented graph theory and used his ideas to show that it is impossible to devise such a route. We shall, in the subsequent sections, devise a broader version of Euler's ideas and explain his solution of the Königsberg bridge problem.

8.2 Fundamental Ideas of Graph Theory

A graph consists of vertices and edges. A graph may be *connected*, that is, consist of one continuous piece or *disconnected*, that is, consist of more than one contiguous piece. See Fig. 8.4. Notice in the figure that the edges of the graph determine certain two-dimensional regions, or faces, in the graph. The graph on the left defines 3 faces, and the graph on the right defines 2 faces. It is customary in this subject to think of the graph as living on a *sphere* rather than in the plane, so that the exterior region (see Fig. 8.5) counts as a face.

Euler's first fundamental insight about graphs is the following theorem:

Theorem 8.1 Let $\mathcal{G}$ be any connected graph on the sphere. Let V be the number of vertices, E the number of edges, and F the number of two-dimensional regions (or faces) defined by the edges. Then

$$V - E + F = 2$$

We should like to spend some time explaining why this important theorem is true. Begin with the simplest possible graph—see Fig. 8.6. It has just one vertex. There are no edges. And there is one face—which is the entire region of the sphere

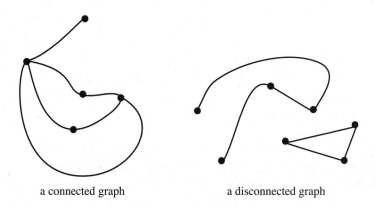

a connected graph   a disconnected graph

Figure 8.4   Connected and disconnected graphs.

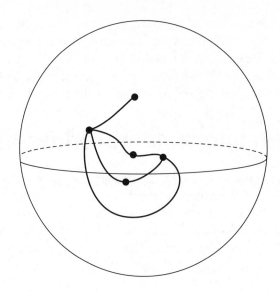

Figure 8.5 A graph on the sphere.

complementary to the single vertex. Thus $V = 1$, $E = 0$, and $F = 1$. Thus we see that

$$V - E + F = 1 - 0 + 1 = 2$$

So Euler's theorem is true in this very simple case.

Now imagine making the graph more complex. We add a single edge, as shown in Fig. 8.7. How have the numbers changed? Well, now $E = 1$. But there is an additional vertex (that is, there is a vertex at each end of the edge), so $V = 2$. And there is still a single face, so $F = 1$. Now

$$V - E + F = 2 - 1 + 1 = 2$$

Thus Euler's theorem remains true.

Now the fundamental insight is that we can build up *any* graph by adding one edge at a time. And there are only three ways that we may add an edge. Let us discuss them one at a time:

- We can add a new edge so that it has one vertex on the existing graph and one vertex free—see Fig. 8.8. In doing so, we add one vertices, add one edge, and do not change the number of faces. Thus, in the formula $V - E + F$, we

●

Figure 8.6 Beginning of the proof of Euler's formula.

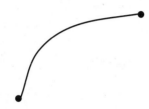

Figure 8.7   Euler's formula for a more complicated graph.

have increased E by 1 and increased V by 1. These two increments cancel out, so the sum of 2 remains unchanged.

- We can add the new edge so that both ends (the vertices) are at the same point on the existing graph—see Fig. 8.9. Thus we have added one edge, no vertices, and one face. As a result, in the formula $V - E + F$, we have increased E by 1 and increased F by 1. These two increments cancel out, so the sum of 2 remains unchanged.

- We can add the new edge so that the two ends are at two different vertices of the existing graph—see Fig. 8.10. so we have added one edge and one face, but no vertices. As a consequence, in the formula $V - E + F$, we have increased E by 1 and increased F by 1. But there are no new vertices. Therefore the two increments cancel out, and the sum of 2 remains unchanged.

This exhausts all the cases, and shows that, as we build up any graph, the Euler sum $V - E + F$ will always be 2.

We call 2 the *Euler characteristic* of the sphere. It is a fundamental geometric invariant of this surface. It turns out that the Euler characteristic of a torus—see Fig. 8.11—is *not* 2. It is in fact 0, as the figure indicates.

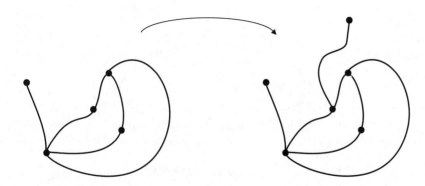

Figure 8.8   The inductive step in the proof of Euler's formula.

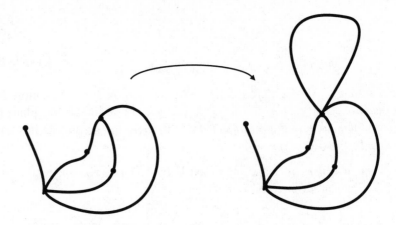

Figure 8.9   The inductive step in the proof of Euler's formula.

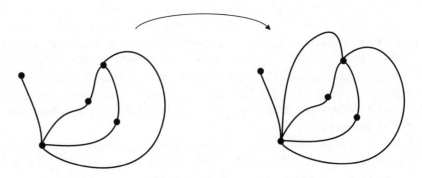

Figure 8.10   The inductive step in the proof of Euler's formula.

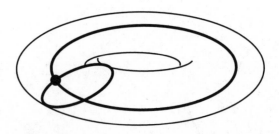

Figure 8.11   The Euler characteristic of a torus.

8.3 Application to the Königsberg
Bridge Problem

Before returning to Euler's original problem, let us look at an even more fundamental question. Let $\{v_1, v_2, \ldots, v_k\}$ be a collection of vertices in the plane (or on
the sphere). The *complete graph* on these vertices is the graph that has an edge
connecting any two of the vertices. As an instance, the complete graph on three
vertices is shown in Fig. 8.12. The complete graph on four vertices is shown in
Fig. 8.13. We may ask whether the complete graph on five vertices can be drawn
in the plane—or on the sphere—without any edges crossing any others. If that
were possible, then the resulting graph would have five vertices (so $V = 5$) and
$\binom{5}{2} = 10$ edges (so $E = 10$) and $\binom{5}{3} = 10$ faces (because every face would have
to be a triangle). But then $V - E + F = 5 - 10 + 10 = 5$, and that is impossible.
The answer is supposed to be 2! We say that the complete graph on five vertices
cannot be imbedded in the plane (or in the sphere). This simple example already
illustrates the power of Euler's formula.

Now let us examine the seven bridges of Königsberg. In Fig. 8.14 we convert the
original Königsberg bridge configuration into a planar graph. We do so by planting
a flag in each land mass defined by the river (there are four land masses in Fig. 8.3)
and connecting two flags if there is a bridge between the two corresponding land
masses. If the *order* of a vertex in a graph is the number of edges that meets at that
vertex, then we see that the graph in Fig. 8.14 has three vertices of order three. This
fact has the following significance.

Imagine a traveler endeavoring to satisfy the stipulations of the Königsberg
bridge problem. If this traveler enters a vertex of order three, then that traveler will
leave that vertex on a different edge (since the traveler is not allowed to traverse
the same edge twice in this journey). But then, if the traveler ever enters that vertex
again, he/she cannot leave. There is no edge left on which the traveler can leave
(since there are only three edges total at that vertex). So the journey would have

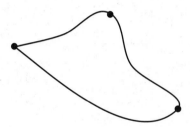

Figure 8.12   The complete graph on three vertices.

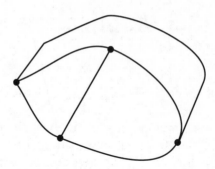

Figure 8.13   The complete graph on four vertices.

to end at that vertex. That is OK, but there are three vertices of order three. The journey cannot end at all three of those vertices—just at one of them. This is a contradiction.

We see therefore, by Euler's original analysis, that it is impossible to find a journey that traverses all seven bridges while not repeating any part of the path.

You Try It: Remove one of the seven bridges from the Pregel River. How does this affect the Königsberg bridge problem? Is it now possible to chart a path, never repeating any part of the route and crossing each bridge precisely once? Does it matter which one of the seven bridges you remove?

Let us say that a graph has an *Euler path* if it traces each edge once and only once. We have shown that the graph corresponding to the original seven Königsberg bridges does *not* have an Euler path.

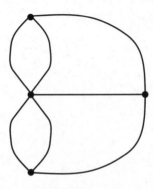

Figure 8.14   The graph corresponding to the Königsberg bridge configuration.

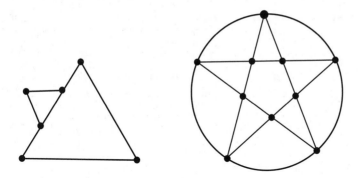

Figure 8.15 Two graphs for Euler analysis.

In a given graph, call a vertex *odd* if an odd number of edges meet at that vertex. Call the vertex *even* if an even number of edges meet at that vertex.

You Try It: Explain why, if a graph has more than two odd vertices, then it does not have an Euler path.

You Try It: Examine each of the graphs in Fig. 8.15. Which of these has an Euler path? If it does, then find this path.

You Try It: Examine each of the graphs in Fig. 8.16. Which of these has an Euler path? If it does, then find this path.

You Try It: Examine each of the graphs in Fig. 8.17. Which of these has an Euler path? If it does, then find this path.

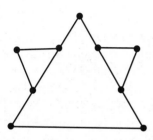
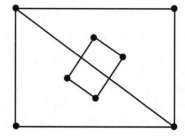

Figure 8.16 Two more graphs for Euler analysis.

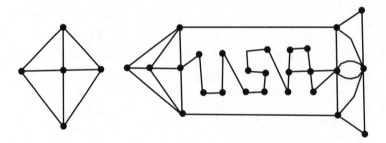

Figure 8.17   Yet two more graphs for Euler analysis.

8.4 Coloring Problems

Many mathematic problems originate among professional mathematicians at universities. After all, they are the folks who spend all day every day thinking about mathematics. They are well qualified to identify and develop interesting directions to investigate. But it also happens that some fascinating and long-standing mathematics problems will originate with laymen. The celebrated four-color problem is an example of such.

In 1852 Francis W. Guthrie, a graduate of University College London, posed the following question to his brother Frederick:

> Imagine a geographic map on the earth (that is, a sphere) consisting of countries only—no oceans, lakes, rivers, or other bodies of water. The only rule is that a country must be a single contiguous mass—in one piece, and with no holes—see Fig. 8.18. As cartographers, we wish to *color* the map so that no two adjacent countries will be of the same color (Fig. 8.19—note that R, G, B, and Y stand for red, green, blue, and yellow). How many colors should the map-maker keep in stock so that he can be sure he can color any map?

Frederick Guthrie was a student of Augustus De Morgan (1806–1871), and ultimately communicated the problem to his mentor. The problem was passed around among academic mathematicians for a number of years [in fact De Morgan communicated the problem to William Rowan Hamilton (1805–1865)]. The first allusion in print to the problem was by Arthur Cayley (1821–1895) in 1878.

The eminent mathematician Felix Klein (1849–1925) in Göttingen heard of the problem and declared that the only reason the problem had never been solved is that no capable mathematician had ever worked on it. He, Felix Klein, would offer a class, the culmination of which would be a solution of the problem. He failed.

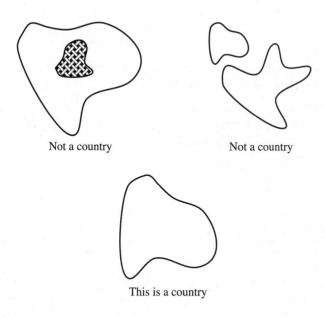

Not a country   Not a country

This is a country

Figure 8.18   Map coloring.

In 1879, A. Kempe (1845–1922) published a solution of the four-color problem. That is to say, he showed that any map whatever could be colored with four colors. Kempe's proof stood for 11 years. Then a mistake was discovered by P. Heawood (1861–1955). Heawood studied the problem further and came to a number of fascinating conclusions:

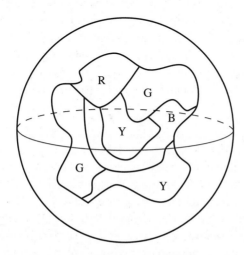

Figure 8.19   The four-color problem.

- Kempe's proof, particularly his device of "Kempe chains," *does* suffice to show that any map whatever can be colored with five colors.

- Heawood showed that if the number of edges around each region in the map is divisible by three, then the map is four-colorable.

- Heawood found a formula that gives an estimate for the "chromatic number" of any surface. Here the chromatic number $\chi(g)$ of a surface is the least number of colors it will take to color *any* map on that surface. We write the chromatic number as $\chi(g)$. In fact the formula is

$$\chi(g) \le \left\lfloor \frac{1}{2}\left(7 + \sqrt{48g + 1}\right) \right\rfloor$$

so long as $g \ge 1$.

Here is how to read this formula. It is known, thanks to work of Camille Jordan (1838–1922) and August Möbius (1790–1868), that any surface in space is a sphere with handles attached (see Fig. 8.20). The number of handles is called the *genus*, and we denote it by g. The Greek letter chi (χ) is the chromatic number of the surface—the least number of colors that it will take to color any map on the surface. Thus $\chi(g)$ is the number of colors that it will take to color any map on a surface that consists of the sphere with g handles. Next, the symbols $\lfloor \ \rfloor$ stand for the "greatest integer function." For example $\lfloor \frac{9}{2} \rfloor = 4$ just because the greatest integer in the number "four and a half" is 4. Also $\lfloor \pi \rfloor = 3$ because $\pi = 3.14159\ldots$ and the greatest integer in the number pi is 3.

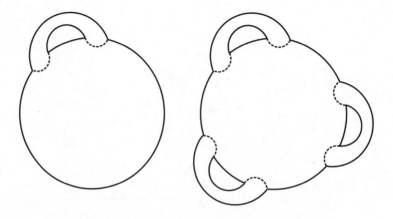

Figure 8.20   The structure of a closed surface in space.

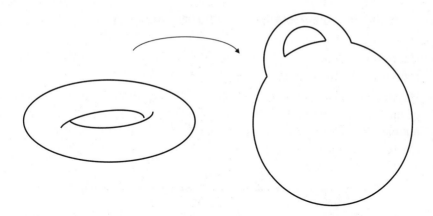

Figure 8.21   The torus is a sphere with one handle.

Now a sphere is a sphere with no handles, so $g = 0$. We may calculate that

$$\chi(g) \leq \left\lfloor \frac{1}{2}\left(7 + \sqrt{48 \cdot 0 + 1}\right)\right\rfloor = \left\lfloor \frac{1}{2}(8)\right\rfloor = 4$$

This is the four-color theorem! Unfortunately, Heawood's proof was only valid when the genus is at least 1. It gives no information about the sphere.

The torus (see Fig. 8.21) is topologically equivalent to a sphere with one handle. Thus the torus has genus $g = 1$. Then Heawood's formula gives the estimate 7 for the chromatic number. And in fact we can give an example of a map on the torus that requires seven colors. Here is what Fig. 8.22 shows. It is convenient to take a pair of scissors and cut the torus apart. With one cut, the torus becomes a cylinder; with the second cut it becomes a rectangle.

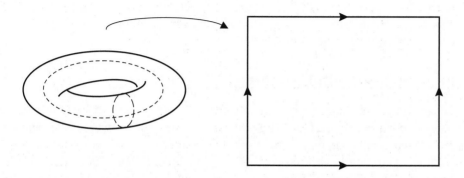

Figure 8.22   The torus as a rectangle with identifications.

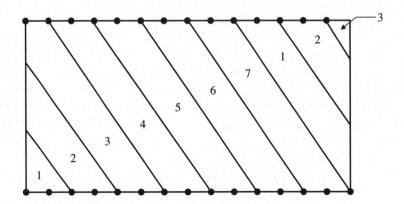

Figure 8.23   A map on the torus that requires seven colors.

The arrows on the edges indicate that the left and right edges are to be identified (with the same orientation), and the upper and lower edges are to be identified (with the same orientation). We call our colors "1," "2," "3," "4," "5," "6," "7." The reader may verify that there are seven countries shown in our Fig. 8.23, and every country is adjacent to (that is, touches) every other. Thus they all must have different colors! This is a map on the torus that requires seven colors; it shows that Heawood's estimate is sharp for this surface.

Heawood was unable to decide whether the chromatic number of the sphere is 4 or 5. He was also unable to determine whether any of his estimates for the chromatic numbers of various surfaces of genus $g \geq 1$ were sharp or accurate. That is to say, for the torus (the closed surface of genus 1), Heawood's formula says that the chromatic number does not exceed 7. Is that in fact the best number? Is there a map on the torus that really requires seven colors? And for the torus with two handles (genus 2), Heawood's estimate gives an estimate of 8. Is that the best number? Is there a map on the double torus that actually *requires* eight colors? And so forth: we can ask the same question for every surface of every genus. Heawood could not answer these questions.

8.4.1 MODERN DEVELOPMENTS

The late nineteenth century saw more alleged solutions of the four-color problems, many of which stood for as long as 11 years. Eventually errors were found, and the problem remained open on into the twentieth century.

What is particularly striking is that Gerhard Ringel (1919– 2008) and J. W. T. Youngs (1910–1970) were able to prove in 1968 that all of Heawood's estimates, for the chromatic number of any surface of genus at least 1, are sharp. So the chromatic number of a torus is indeed 7. The chromatic number of a "double-torus" with two holes is 8. And so forth. But the Ringel/Youngs proof, just like the Heawood formula, does not apply to the sphere. They could not improve on Heawood's result that five colors will always suffice. The four-color problem remained unsolved.

Then in 1974 there was blockbuster news. Using 1200 hours of computer time on the University of Illinois supercomputer, Kenneth Appel and Wolfgang Haken showed that in fact four colors will always work to color any map on the sphere. Their technique is to identify 633 fundamental configurations of maps (to which all others can be reduced) and to prove that each of them is reducible to a simpler configuration. But the number of "fundamental configurations" was very large, and the number of reductions required was beyond the ability of any human to count. And the reasoning is extremely intricate and complicated. Enter the computer.

In those days computing time was expensive and not readily available, and Appel and Haken certainly could not get a 1200-hour contiguous time slice for their work. So the calculations were done late at night, "off the record," during various down times. In fact, Appel and Haken did not know for certain whether the calculation would ever cease. Their point of view was this:

1. If the computer finally stopped then it will have checked all the cases and the four-color problem was solved.
2. If the computer never stopped then they could draw no conclusion.

Well, the computer stopped. But the level of discussion and gossip and disagreement in the mathematical community did not. Was this really a proof? The computer had performed tens of millions of calculations. Nobody could ever check them all.

But now the plot thickens. Because in 1975 a mistake was found in the proof. Specifically, there was something amiss with the algorithm that Appel and Haken fed into the computer. It was later repaired. The paper was published in 1976. The four-color problem was declared to be solved.

In a 1986 article, Appel and Haken point out that the reader of their seminal 1976 article must face

1. 50 pages containing text and diagrams;
2. 85 pages filled with almost 2500 additional diagrams;
3. 400 microfiche pages that contain further diagrams and thousands of individual verifications of claims made in the 24 statements in the main section of the text.

But it seems as though there is always trouble in paradise. Errors continued to be discovered in the Appel/Haken proof. Invariably the errors were fixed. But the stream of errors never seemed to cease. So is the Appel/Haken work really a proof?

Well, there is hardly anything more reassuring than another, independent proof. Paul Seymour and his group at Princeton University found another way to attack the problem. In fact they found a new algorithm that seems to be more stable. They also needed to rely on computer assistance. But by the time they did their work computers were *much, much* faster. So they required much less computer time. In any event, this paper appeared in 1994.

8.4.2 DENOUEMENT

It is still the case that mathematicians are most familiar with, and most comfortable with, a traditional, self-contained proof that consists of a sequence of logical steps recorded on a piece of paper. We still hope that some day there will be such a proof of the four-color theorem. After all, it is only a traditional, Euclidean-style proof that offers the understanding, the insight, and the sense of completion that all scholars seek.

And there are new societal needs: theoretical computer science and engineering and even modern applied mathematics require certain pieces of information and certain techniques. The need for a workable device often far exceeds the need to be *certain* that the technique can stand up to the rigorous rules of logic. The result may be that we shall reevaluate the foundations of our subject. The way that mathematics is practiced in the year 2100 may be quite different from the way that it is practiced today.

8.5 The Traveling Salesman Problem

It is a charming fact of life that some of the most fascinating mathematical problems have utterly simple statements that can be understood by most anyone. Fermat's last theorem is such a problem. The four-color problem (see Sec. 8.4) is another. A problem that is fairly old, and is still of preeminent importance both for mathematics and logic and also for theoretical computer science, is the celebrated "traveling salesman problem" (TSP). We shall discuss this problem in the present section.

First studied in the mid-nineteenth century by William Rowan Hamilton (1805–1865) and Thomas Kirkman (1806–1895), the question concerns traveling a circuit in the most efficient fashion. The question is often formulated in terms of a traveling salesman who must visit cities $C_1, C_2, \ldots, C_k$. There is a path connecting any city to any other, and a cost assigned to each path. The goal of the salesman is to begin at some city—say C_1—and to visit every city precisely once. The trip is to end again at C_1. And obviously the salesman wants to minimize his cost. See Fig. 8.24.

We may use our knowledge of counting techniques to quickly get an estimate of the number of possible paths that the salesman can take. For the first leg of the

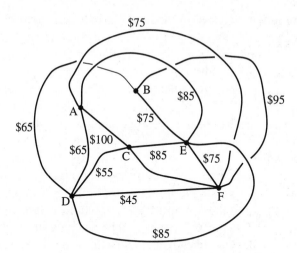

Figure 8.24 The traveling salesman problem.

trip, the salesman (beginning at C_1) may choose to go to C_2 or C_3 or ... C_k. Thus there are $(k-1)$ choices. For the next leg, the salesman may choose any of the remaining $(k-2)$ cities. So there are $(k-2)$ choices. And so forth. In summary, there are

$$(k-1) \cdot (k-2) \cdots 3 \cdot 2 \cdot 1 = (k-1)!$$

possible paths that the traveling salesman might take. According to a formula of Stirling,

$$(k-1)! \approx \sqrt{2\pi(k-1)} \cdot \frac{(k-1)^{k-1}}{e^{k-1}} \tag{8.1}$$

In particular, the number is exponential in k. This means that the problem is difficult, and takes a great many steps.

A variety of techniques are known for *estimating* the correct solution to the traveling salesman problem. In particular, if one is willing to settle for a path that cost *not more than twice* the optimal path, then one may find a solution rather efficiently. But the truly optimal solution can be quite complex to find. As an instance, in 2001 the optimal tour of 15,112 German cities was found—and it was shown that the given solution was indeed optimal. The calculation was performed on a 110-CPU parallel-processing computer and required 22.6 years of computer time on a single 500 MHz Alpha processor. In 2006 the optimal circuit for 85,900 cities was found. It took over 136 computing years on a CONCORDE chip.

A current and timely instance of the traveling salesman problem occurs in the manufacture of electronic circuits. These days such circuits are solid state, and all the components are mounted on a board. There are often thousands of units, and much of the manufacture is automated. Certainly part of the process is that a drill head must travel around the circuit board making holes for the various electronic components. Since many tens of thousands of circuit boards will be manufactured, one wants the tour of the drill head to be as efficient as possible.

It has been argued that the single most important problem today in the mathematical sciences is the **P** vs. **NP** problem. Roughly speaking, this is a problem of considering when a problem can be solved in polynomial time and when it will take exponential time. As a simple instance, the problem of taking N randomly shuffled cards and putting them in order is of polynomial time because one first goes through all N cards to find the first card, then one looks through the remaining $(N - 1)$ cards to find the second card, and so forth. In short, it takes at most

$$N + (N - 1) + (N - 2) + \cdots 3 + 2 + 1 = \frac{N(N + 1)}{2} < N^2$$

steps to sort the cards. This is a polynomial estimate on the number of steps. By contrast, the traveling salesman problem with N cities takes about $(N - 1)! \approx (N/e)^{N-1}$ steps, and hence is of exponential complexity. It is known that a complete solution to the traveling salesman problem is logically equivalent to a solution of the **P/NP** problem. This is one of the Clay Mathematics Institute's 1 million dollar Millenium Prize Problems!

We conclude with a few words about how one might find an efficient (though not necessarily optimal) circuit for the traveling salesman problem. There is a commonly used algorithm, both in tree theory and in graph theory, for finding optimal paths and circuits. It is called the *greedy algorithm*. The idea is simple and intuitive. Suppose we are given a layout of cities and paths connecting them, with a cost connected with each path—see Fig. 8.24. Notice in the figure that every city (node) is connected to every other city. We begin our circuit at the node C with the cheapest path between two cities. In the figure that is path from C to D that costs $55. So, for our first step, we pass from C to D. The next arc must begin at D. We choose the cheapest remaining arc emanating from D (but of course *not* the one we have already traversed). In the figure, that takes us to F over the arc that costs $45. And we continue in this fashion, at every step choosing the cheapest arc possible. This is the greedy algorithm.

There is no guarantee that the greedy algorithm produces the truly *optimal* circuit for the traveling salesman. But it is a theorem that it produces a circuit that costs no more than twice as much as the optimal amount. For many applications this is adequate.

Exercises

1. Give an example of a graph on five vertices without an Euler path.

2. Give an example of a graph on five vertices with two distinct Euler paths.

3. Imagine a torus with two handles. What would be the correct Euler formula for this surface? It should have the form

$$V - E + F = \text{(some number)}$$

   What is that number? The number χ on the right hand side is called the *Euler characteristic* of the surface.

4. Consider the complete graph on six vertices. How many edges does it have? How many faces?

5. How many edges does the complete graph on k vertices have?

6. Consider a graph built on two rows of three vertices for a total of six vertices. Construct a graph by connecting every vertex in the first row to every vertex of the second row and vice versa. How many edges does this graph have?

7. Consider the standard picture of a five-pointed star. This can be thought of as a graph. How many vertices does it have? How many edges?

8. Give an example of a graph with more vertices than edges. Give an example of a graph with more edges than vertices.

9. If a surface can be described as a "sphere with g handles," then we say it has genus g. Thus a lone sphere has genus 0, a torus has genus 1, and so forth. Based on your experience with Exercise 3 above, posit a formula that relates the Euler characteristic χ with the genus g.

CHAPTER 9

Number Theory

9.1 Divisibility

Let $\mathbb{N}$ denote the collection of all positive integers. We call these the *natural numbers*. Number theorists study the properties of $\mathbb{N}$.

Let n and k be two positive integers. We say that k *divides* n if there is another positive integer m such that $n = m \cdot k$. For example, let $n = 12$ and $k = 3$. Certainly 3 divides 12 because

$$12 = 4 \cdot 3$$

On the other hand, 5 does *not* divide 13 because there is no positive integer m with

$$13 = m \cdot 5$$

In fact if we divide 5 into 13 we find that it goes twice with a remainder of 3. We may write

$$13 = 2 \cdot 5 + 3$$

Whenever the remainder is nonzero then the division is not even.

Turning this reasoning around, we may enunciate the Euclidean algorithm. Let n be a positive integer and d another positive integer. Then there are two other positive integers (the *quotient q* and the remainder r) such that

$$n = q \cdot d + r \qquad (9.1)$$

This simple formula—which dates back more than 2000 years and is commonly called the *Eucliean algorithm*—is the basis for our understanding of division. It says that we may divide any integer n by any other integer d. We get an answer, or quotient, q and a remainder r. When the remainder r is equal to 0 then the Eq. (9.1) becomes

$$n = q \cdot d$$

and we see that d divides n.

The Euclidean algorithm has many uses. For example, it is a matter of some interest to find the greatest common divisor of two given positive integers. This can be done by educated guessing. For example, the greatest common divisor of 84 and 18 is 6. But the Euclidean algorithm gives a step-by-step method (that one could program onto a computer) for finding this number.

How does the method work? We divide 18 into 84 with remainder 12:

$$84 = 4 \cdot 18 + 12 \qquad (9.2)$$

Now we divide the remainder 12 into the divisor 18:

$$18 = 1 \cdot 12 + 6 \qquad (9.3)$$

Next we repeat the algorithm by dividing the remainder 6 into the divisor 12:

$$12 = 2 \cdot 6 \qquad (9.4)$$

We see that there is no remainder, so the process must stop. The greatest common divisor is 6 (which is the last divisor that has occurred).

We can also see, by examining this last example, why the method of the Euclidean algorithm must work to find the greatest common divisor. For Eq. (9.4) shows that 6 must divide 12. Then Eq. (9.3) shows that 6 must divide 18. And, finally, Eq. (9.2) shows that 6 must divide 84. So certainly we see that 6 is a common divisor of 18 and 84. Reversing the reasoning, we see that 6 is the greatest number that could have this property.

We say that two positive integers m and n are *relatively prime* if they have no common divisors (except of course 1). If m and n *do* have a common divisor—say *k*—then any prime factor of *k* will also be a common divisor of *m* and *n*. So another way to express the concept is that *m* and *n* are relatively prime if they have no common prime divisors.

This suggests an *ad hoc* way of telling when two given integers are relatively prime. Just write out their prime factorizations and see whether there are any common factors. As an example,

$$90 = 2 \cdot 3^2 \cdot 5$$

and

$$77 = 7 \cdot 11$$

We see explicitly that the numbers 90 and 77 have no common prime factors. So they are relatively prime.

By contrast, consider

$$360 = 2^3 \cdot 3^2 \cdot 5$$

and

$$108 = 2^2 \cdot 3^3$$

We see that these two integers have two factors of 2 in common and two factors of 3 in common. Altogether then, the greatest common divisor of 360 and 108 is $4 \cdot 9 = 2^2 \cdot 3^2 = 36$.

We also know from the discussion above that the Euclidean algorithm may be used to determine the greatest common divisor of two given integers. If that greatest common divisor is 1, then the two integers are relatively prime.

9.2 Primes

The building blocks of the natural numbers are the *prime numbers*. A positive integer is called prime if it is not divisible by any integer except for 1 and itself. The first several primes are 2, 3, 5, 7, 11, 13, 17, 19, 23, 29, 31, ... It is customary *not* to refer to 1 as a prime. The first prime number is 2, and 2 is the only even prime number. It is an old theorem of Euclid that there are in fact infinitely many primes.

The *fundamental theorem of arithmetic* says that every natural number can be written as a product of primes (or powers of primes) in a unique way. For example,

$$2520 = 2^3 \cdot 3^2 \cdot 5 \cdot 7$$

And there is no other way to factor 2520 into primes. Notice that the factors 2 and 3 are repeated—the factor 2 occurs three times and the factor 3 occurs twice. The factors 5 and 7 occur once only.

9.3 Modular Arithmetic

One of the most useful devices for studying the integers is modular arithmetic. We say that two integers m and n are *equivalent modulo k* if the number $m - n$ is divisible by k. This concept is best illustrated with some examples.

EXAMPLE 9.1
Let $k = 2$. The numbers 3 and 5 are equivalent modulo 2 just because $5 - 3 = 2$ is divisible by 2. In the same way, 5 and 11 are equivalent modulo 2 because $11 - 5 = 6$ is divisible by 2. We write

$$5 = 3 \mod 2$$

and

$$11 = 5 \mod 2$$

In fact it turns out that any two odd integers are equivalent modulo 2 because the difference of two odd integers will be an even number.

We also note that 4 and 12 are equivalent modulo 2 because $12 - 4 = 8$ is divisible by 2. In the same way, 24 and 48 are equivalent modulo 2 because $48 - 24 = 24$ is divisible by 2. In point of fact, any two even numbers are equivalent modulo 2. $\square$

EXAMPLE 9.2
Let $k = 12$. The numbers 4 and 18 are *not* equivalent modulo 12 because $18 - 4 = 14$ is not divisible by 12. However, the numbers 13 and 37 *are* equivalent modulo 12 because $37 - 13 = 24$ *is* divisible by 12.

The numbers 125 and 185 are equivalent modulo 12 because $185 - 125 = 60$ is divisible by 12. The numbers 5 and 132 are not equivalent modulo 12. $\square$

Modularity respects the arithmetic operations. For example, if

$$n = m + \ell$$

then

$$n \bmod k = m \bmod k + \ell \bmod k$$

Also, if

$$n = m \cdot \ell$$

then

$$n \bmod k = (m \bmod k) \cdot (\ell \bmod k)$$

As an application of these last ideas, we can see quickly that 1347 does not divide 25168. For if

$$25168 = m \cdot 1347$$

then

$$25168 \bmod 3 = (m \bmod 3) \cdot (1347 \bmod 3)$$

or

$$1 \bmod 3 = (m \bmod 3) \cdot 0 = 0 \bmod 3$$

This is impossible.

9.4 The Concept of a Group

A *group* is a set G, or a collection of objects, together with a binary operation for combining them. We usually denote the binary operation by "$\cdot$". We assume that this binary operation satisfies certain basic and plausible properties:

1. **Associativity** If $g, h, k \in G$ then $g \cdot (h \cdot k) = (g \cdot h) \cdot k$.
2. **Identity element** There is a distinguished element $e \in G$ such that, for all $g \in G$, $e \cdot g = g \cdot e = g$.
3. **Multiplicative inverse** For each $g \in G$ there is an element $h \in G$ such that $g \cdot h = h \cdot g = e$.

Notice that we do *not* assume that a group is commutative; that is, we do not assume that $g \cdot h = h \cdot g$ for all $g, h \in G$. The property of associativity that we postulate in Axiom 1 is a different property: it says that when we are combining three elements we may group them, two by two, in either of the two obvious ways; the same answer results. A group that is commutative is called *abelian* in honor of Niels Henrik Abel (1802–1829).

EXAMPLE 9.3
Let G be the positive real numbers and let the group operation be multiplication: $P(x, y) = x \cdot y$, where $\cdot$ is ordinary multiplication of reals. Then (G, P) is a group.

    Axiom 1: Of course multiplication of real numbers is associative.

    Axiom 2: The number 1 is the identity element for multiplication: $1 \cdot x = x \cdot 1 = x$ for any real number x.

    Axiom 3: The multiplicative inverse of a group element is its ordinary reciprocal. That is, if $x \in \mathbb{R}$ satisfies $x > 0$ then $1/x$ is its multiplicative inverse. □

EXAMPLE 9.4
Let G be the integers and let $P(x, y) = x + y$ (ordinary addition). Then (G, P) is a group.

    Axiom 1: Certainly addition of integers is associative.

    Axiom 2: The number 0 is the additive identity.

    Axiom 3: The additive inverse of a group element is its negative: if $m \in \mathbb{Z}$ then $-m$ is its group inverse. □

EXAMPLE 9.5
Let G be the $k \times k$ matrices with real entries and nonzero determinant. This is sometimes called the *general linear group on k letters* and is denoted by $GL(k, \mathbb{R})$.
    Let P be ordinary matrix multiplication. Then (G, P) is a group.

    Axiom 1: Matrix multiplication is associative.

    Axiom 2: The group identity is the matrix

$$
I_k \equiv \left.\begin{pmatrix}
1 & 0 & 0 & \ldots & 0 \\
0 & 1 & 0 & \ldots & 0 \\
 & & \ldots & & \\
0 & 0 & \ldots & 1 & 0 \\
0 & 0 & \ldots & 0 & 1
\end{pmatrix}\right\} k
$$

$$\underbrace{}_{k}$$

Thus, if $m \in G$, then $I_k \cdot m = m \cdot I_k = m$.

Axiom 3: The multiplicative inverse of a group element is its matrix inverse. Thus if $m \in G$ then the inverse matrix m^{-1} is the group inverse.

Notice in this example that it is important to restrict attention to square matrices, so that multiplication of any two elements in any order will make sense. We also require that each matrix have nonzero determinant, so that each matrix will have an inverse. To see that G is closed under the group operation of matrix multiplication, we must note that if $M, N \in G$ then $\det(M \cdot N) = (\det M)(\det N) \neq 0$. $\quad\square$

Unlike the previous two examples, this last one is a noncommutative group.

The advantage of the axiomatic method, in the present context, is that when we prove a proposition or theorem about "a group G," it applies simultaneously to all groups. Thus the axiomatic method gives us both a way of being concise and a way of cutting to the heart of the matter.

Proposition 9.1 *The multiplicative identity for a group is unique.*

Proof: Let G be a group. Let e and e' both be elements of G that satisfy Axiom 2. Then

$$e = e \cdot e' = e'$$

Thus e and e' must be the same group element. $\quad\square$

Proposition 9.2 *Let G be a group and $g \in G$. Then there is only one multiplicative inverse for g.*

Proof: Suppose that h and k both satisfy the properties of the multiplicative inverse (Axiom 3) relative to g. Then

$$h = h \cdot e = h \cdot (g \cdot k) = (h \cdot g) \cdot k = e \cdot k = k$$

Thus h and k must be the same group element, establishing that the multiplicative inverse is unique. $\quad\square$

Proposition 9.3 *Let g be an element of the group G. Then*

$$(g^{-1})^{-1} = g$$

Proof: Observe that

$$g \cdot g^{-1} = e$$

and

$$g^{-1} \cdot g = e$$

Thus g satisfies the properties of the inverse element (Axiom 3) relative to g^{-1}. Since the last proposition establishes that the inverse element for g is unique, it follows that g must be the multiplicative inverse for g^{-1}. In other words, $(g^{-1})^{-1} = g$. □

Proposition 9.4 *Let g, h be elements of a group G. Then $(g \cdot h)^{-1} = h^{-1} \cdot g^{-1}$.*

Proof: We calculate that

$$
\begin{aligned}
[h^{-1} \cdot g^{-1}] \cdot [g \cdot h] &= h^{-1} \cdot [g^{-1} \cdot [g \cdot h]] \\
&= h^{-1} \cdot [(g^{-1} \cdot g) \cdot h] \\
&= h^{-1} \cdot [e \cdot h] \\
&= h^{-1} \cdot h \\
&= e
\end{aligned}
$$

A similar calculation shows that

$$
[g \cdot h] \cdot \left[h^{-1} \cdot g^{-1}\right] = e
$$

The assertion follows. □

Definition 9.1 Let G be a group and $H \subset G$. We call H a *subgroup* of G if the following properties hold

Closure: The group operation P associated with G satisfies $P : H \times H \to H$. In other words, H is closed under the group operation of G (see the Exercises at the end of Chap. 5 for the concept of "closed");

Identity element: The group identity e is an element of H;

Multiplicative inverse: If $h \in H$ then its group inverse element h^{-1} lies in H.

Notice that the point of the last definition is that H is itself a group, using operations (and the group identity) inherited from the larger group G.

EXAMPLE 9.6
The pair $(\mathbb{Q}, +)$ of the rational numbers under the ordinary operation of addition forms a group. Then the integers $\mathbb{Z} \subset \mathbb{Q}$ form a subgroup. That is, the integers are a group under the same operation of addition. They are closed under this operation. □

EXAMPLE 9.7

Let G be the 3×3 matrices with real entries and nonvanishing determinant

$$\begin{pmatrix} a & b & c \\ d & e & f \\ g & h & i \end{pmatrix}$$

Let the group law be matrix multiplication. As we noted in Example 9.4, this is a group. Let H be the subset of G consisting of those matrices with nonzero entries on the main diagonal and zero entries off the diagonal.

$$\begin{pmatrix} a & 0 & 0 \\ 0 & e & 0 \\ 0 & 0 & i \end{pmatrix}$$

Then H is a subgroup.   □

Let G be a group and H a subgroup. Let us define a relation $\mathcal{R}$ on G as follows: $(x, y) \in \mathcal{R}$ provided that $x^{-1} \cdot y \in H$. If x is any element of G then of course $x^{-1} \cdot x = e \in H$ so the relation is reflexive. Notice that $x^{-1}y \in H$ if and only if $y^{-1}x \in H$, for these elements are multiplicative inverses of each other, and H is a group. Thus the relation is symmetric. Finally, suppose that $x, y, z \in G$, that x is related to y, and that y is related to z. Then $x^{-1} \cdot y \in H$ and $y^{-1} \cdot z \in H$. As a result, $[x^{-1} \cdot y] \cdot [y^{-1} \cdot z] = x^{-1} \cdot z \in H$. So we may conclude that x is related to z, and the relation is transitive.

Therefore we have an equivalence relation. The equivalence relation partitions the group G into pairwise disjoint subsets. What do these subsets look like?

If $x \in G$ then define $xH = \{x \cdot h : h \in H\}$. The set xH is called a *coset* of H. Notice the following properties:

(1) **Elements of xH are distinct:** If $h, k \in H$ then $xh = xk$ implies $x^{-1}(xh) = x^{-1}(xk)$ hence $h = k$. So the elements of xH are distinct.

(2) **If $a, b \in xH$ then $a\mathcal{R}b$:** If $a, b \in xH$ then $a = xh$ and $b = xk$ for some $h, k \in H$. But then

$$a^{-1}b = (xh)^{-1}(xk) = h^{-1}x^{-1}xk = h^{-1}k \in H$$

Thus a and b are related under $\mathcal{R}$.

(3) **If $a \in xH$ and $a\mathcal{R}b$ then $b \in xH$:** Now let $a \in xH$ and assume that $(a, b) \in \mathcal{R}$. Thus $a^{-1}b \in H$. So there is an element $h \in H$ such that $a = xh$ and there is another element $k \in H$ such that $a^{-1}b = k$. But then $b = ak = (xh)k = x(hk) \in xH$.

It follows from (2) and (3) that the equivalence classes induced by $\mathcal{R}$ are precisely the cosets of H. It follows from (1) that, if H has finitely many elements, then each coset has the same number of elements.

We let G/H denote the collection of cosets of H in G. With an additional condition on H (that H be a *normal subgroup*), G/H can actually be made into a group. We shall not explore that idea here.

Theorem 9.1 *Let $G = \{g_1, \ldots, g_k\}$ be a group with finitely many elements. Let $H \subset G$ be a subgroup with m elements. Then the integer m evenly divides the integer k.*

Proof: The group G partitions into the cosets of H. Each coset has m elements, and the cosets are of course pairwise disjoint. That means that m divides k.

EXAMPLE 9.8
Let a relation on the integers $\mathbb{Z}$ be defined by $x \mathcal{R} y$ if $y - x$ is evenly divisible by 6. This is an equivalence relation. There are six equivalence classes, namely

$$E_0, E_1, E_2, E_3, E_4, E_5$$

Indeed,

$$E_0 = \{\ldots, -12, -6, 0, 6, 12, \ldots\}$$
$$E_1 = \{\ldots, -11, -5, 1, 7, 13, \ldots\}$$
$$E_2 = \{\ldots, -10, -4, 2, 8, 14, \ldots\}$$
$$E_3 = \{\ldots, -9, -3, 3, 9, 15, \ldots\}$$
$$E_1 = \{\ldots, -8, -2, 4, 10, 16, \ldots\}$$
$$E_2 = \{\ldots, -7, -1, 5, 11, 17, \ldots\}$$

and so forth.

We add two equivalence classes as follows:

$$E_j + E_k = E_{j+k}$$

For instance

$$E_3 + E_4 = E_7 = E_1$$

You should check that this notion of addition is well defined (unambiguous). Also, the identity element is E_0 and each E_m has E_{-m} as its additive inverse. In sum, the collection of equivalence classes forms a group. This group is usually denoted by $\mathbb{Z}/6\mathbb{Z}$ or $\mathbb{Z}_6$ or $\mathbb{Z}/6$. It is a group having six elements; we call this a group of *order* 6. In general, the order of a group with finitely many elements is just the number of elements in the group.

Let H be the subset of $\mathbb{Z}_6$ consisting of E_0, E_2, E_4. Verify that this is a subgroup of order 3. Notice that 3 divides 6. The only other nontrivial subgroup is that consisting of E_0 and E_3. Notice that it has order 2.

The number 6 has no nontrivial divisors besides 2 and 3 and the group $\mathbb{Z}/6\mathbb{Z}$ has no other nontrivial subgroups.   □

Of course there is nothing special about the number 6 in the last example. If k is any positive integer then we may declare that $x \mathcal{R} y$ if $y - x$ is evenly divisible by k. The result is a group of order k denoted by $\mathbb{Z}/k\mathbb{Z}$ or $\mathbb{Z}_k$ or $\mathbb{Z}/k$. If the positive integer m evenly divides k ($k = m \cdot p$) then there will be one subgroup of order m and that group will consist of the elements $E_0, E_p, E_{2p}, \ldots, E_{(m-1)p}$. When the context is understood we write the elements of $\mathbb{Z}_k$ as $0, 1, 2, \ldots, k - 1$. We say that we are doing *arithmetic modulo k* or *arithmetic mod k*.

EXAMPLE 9.9

Consider the set $\mathbb{Z}_4 \times \mathbb{Z}_4$. This set may be conveniently thought of as the set of ordered pairs (x, y) where $x, y \in \mathbb{Z}_4$. Define

$$(x, y) + (x', y') = (x + x', y + y')$$

where the addition is performed according to the group law of $\mathbb{Z}_4$. Then $G = \mathbb{Z}_4 \times \mathbb{Z}_4$ so equipped is a group of order 16.

The number 4 divides the order of G, but now there is more than one subgroup having order (that is, number of elements) 4. One such subgroup is $H = \{(0, 0), (1, 0), (2, 0), (3, 0)\}$. Another is $K = \{(0, 0), (0, 1), (0, 2), (0, 3)\}$. Yet another is $L = \{(0, 0), (2, 0), (0, 2), (2, 2)\}$.   □

Now let G be a group of finite order. Let g be a fixed element of G. Consider the set H of all "powers" of g: $g^1 = g, g^2 = g \cdot g, g^3 = g \cdot g \cdot g, \ldots$ as well as $g^{-1}, g^{-2} \equiv (g^{-1})^2, g^{-3} \equiv (g^{-1})^3, \ldots$ Of course $g^0 = e$. It is easy to see that H is a subgroup of G. We say that H is a *cyclic* group (subgroup) because it consists of powers of the single element g.

Let k be the order of H. Then k will be the least positive integer such that $g^k = e$. Since H is a subgroup of G, we see that $k = \text{order } H$ must evenly divide $m = \text{order } G$. It follows (provide the details as an exercise) that $g^m = e$. This conclusion is so important that we display it in a theorem.

Theorem 9.2 *Let G be a group of finite order m. If $g \in G$ then $g^m = e$.*

The examples we have presented raise a natural question. If G is a group of order m and if k evenly divides m then does it follow that G has a subgroup of order k? In general the answer is "no." You are requested in the exercises to provide a counterexample. However, the following theorem of Sylow provides a positive answer in a large number of important instances.

Theorem 9.3 (Sylow) *Let G be a group of finite order and let p be a prime. Suppose that j is a positive integer and that p^j evenly divides the order of G. Then G has a subgroup of order p^j. Indeed, it has subgroups of all orders $p^\ell, 0 \le \ell \le j$.*

Proof of a Special Case of the Sylow Theorems: Let G be a finite group and suppose that the prime integer p divides the order n of G. We shall show that G has a subgroup of order p.

First suppose that the integer m has the property that $g^m = e$ for every $g \in G$ (we call m an *exponent* for G). Let $b \in G$, $b \ne e$, and let H be the cyclic group generated by b (that is, the collection of all powers of b). It can be checked that G/H forms a group. Then of course $b^m = 1$ hence m is an exponent for G/H. Thus the order of G/H divides a power of m. Because

$$[G : e] = [G : H] \cdot [H : e]$$

we may conclude therefore that the order of G divides a power of m.

Since p divides n, there an element $x \in G$ such that the period of x (that is, the minimal power of x that equals e) is divisible by p. Let the period be ps for some integer s. Then $x^s \ne e$ and x^s has period p. Thus x^s generates a subgroup of order p. That is what was to be shown. $\square$

We refer the interested reader to [LAN] or [HER] for a complete consideration of the Sylow theorems and their proofs.

We close with a few remarks about the concept of isomorphism of groups. Consider first an example. You are familiar with the group $\mathbb{Z}_2$. It is a group of order 2. Now suppose we consider a group G with two elements: $G = \{e, m\}$. The element e will be the group identity and the rules of multiplication are

$$e \cdot m = m \qquad m \cdot e = m \qquad e \cdot e = e \qquad m \cdot m = e$$

You can check for yourself that, with this binary operation, G is indeed a group—that is, it satisfies the axioms for a group.

On a formal level, the group $\mathbb{Z}_2$ and the group G are *different*. One has elements $0 \bmod 2$, $1 \bmod 2$ and the other has elements e, m. But in fact it turns out that they are the *same* group; they differ only in the sense that different names have been given to the elements. To see this, write out the group law for $\mathbb{Z}_2$:

$$0+1=1 \qquad 1+0=1 \qquad 0+0=0 \qquad 1+1=0$$

Actually do this on a piece of paper. Now erase each occurrence of $+$ and replace it with $\cdot$. Also erase each occurrence of 0 and replace it by e and erase each occurrence of 1 and replace it by m. What results is the group law that we specified for the group G. This shows that $\mathbb{Z}_2$ and G are precisely the same group, with just different names for the elements and for the binary operation.

We have just seen an example of group isomorphism. Now we can give a formal definition.

Definition 9.2 Let G and H be groups. A function $\phi : G \rightarrow H$ is said to be a *group isomorphism* if it has the following properties

1. The function ϕ is one-to-one and onto.
2. If $g_1, g_2 \in G$ then $\phi(g_1 \cdot g_2) = \phi(g_1) \cdot \phi(g_2)$.

We say that G and H are *isomorphic* and we call the function ϕ an *isomorphism*.

It is the second condition of the definition that says, in effect, that both G and H have the same group law. Let us derive a few simple properties of group isomorphisms.

Proposition 9.5 *Let $\phi : G \rightarrow H$ be an isomorphism of groups. If e_G is the group identity in G and e_H is the group identity in H then $\phi(e_G) = e_H$.*

Proof: We calculate that

$$\phi(e_G) = \phi(e_G \cdot e_G) = \phi(e_G) \cdot \phi(e_G)$$

The expressions that appear on the left and on the far right of this equation are elements of H. Multiplying both sides of this equation on the right by $[\phi(e_G)]^{-1}$ we find that

$$e_H = \phi(e_G)$$

That is what we wished to prove.   □

Proposition 9.6 *Let $\phi : G \to H$ be a group isomorphism. If $g \in G$ then the group inverse, in the group H, of $\phi(g)$ is $\phi(g^{-1})$.*

Proof: We may check that

$$\phi(g) \cdot \phi(g^{-1}) = \phi(g \cdot g^{-1}) = \phi(e_G) = e_H \qquad \text{(by Proposition 9.5)}$$

Also

$$\phi(g^{-1}) \cdot \phi(g) = \phi(g^{-1} \cdot g) = \phi(e_G) = e_H \qquad \text{(by Proposition 9.5)}$$

Thus $\phi(g^{-1})$ possesses the defining properties of the group inverse of $\phi(g)$. Since the group inverse of any group element is unique, our result follows. $\square$

The theory of groups has become a large and essential part of modern mathematics. It is also used in physics (in quantum mechanics, for instance), in engineering, and in theoretical computer science (for example, data compression theory uses group theory).

It is a classical result of basic group theory that all finite abelian groups have been classified. Indeed, it can be shown that any such group is a product (in the sense of set theory) of cyclic groups. One of the triumphs of twentieth century mathematics is that all groups of finite order have been classified. This result is the product of the work of hundreds of mathematicians and will ultimately produce a book of several thousand pages.

9.5 Some Theorems of Fermat

Pierre de Fermat (1601–1665) was one of the most remarkable mathematicians in history. A judge in Toulouse by profession, Fermat studied mathematics as an amateur on his own time. Yet he became one of the most prominent mathematicians of his day, corresponding with many of the great professors throughout Europe. And he is remembered for seminal contributions to mathematics.

One of Fermat's basic results—which is widely used today in cryptography and theoretical computer science, is known as *Fermat's little theorem*:

Theorem 9.4 *Let p be a prime and let a be any positive integer. Then p divides $a^p - a$.*

We may state Fermat's little theorem more succinctly, using the language of modular arithmetic, as

$$a^p \bmod p = a \bmod p$$

Yet another variant is that if a is an integer that is relatively prime to p then

$$a^{p-1} = a \bmod p$$

This last formulation is quickly checked using group theory. For the set of numbers relatively prime to p forms a group under multiplication, and there are $p - 1$ of them. So any number relatively prime to p raised to the power which is the order of the group will give the identity element, or 1. That is the result.

One can verify Fermat's result by using proof by induction on a. We omit the details, but refer the reader to [HER].

Exercises

1. Calculate each of the following modular quantities:

   (a) $3 + 9 \bmod 5$
   (b) $4 \cdot 6 \bmod 3$
   (c) $9 - 4 \bmod 2$
   (d) $2 \cdot 5 \bmod 4$
   (e) $2 + 6 \bmod 3$
   (f) $4 - 9 \bmod 2$

2. If m and n are positive integers then explain why

$$(m \bmod 2) \cdot (n \bmod 2) = (m \cdot n) \bmod 2$$

3. If m and n are positive integers then explain why

$$(m \bmod 2) + (n \bmod 2) = (m + n) \bmod 2$$

4. Show that 111 and 211 are relatively prime.
5. Find the greatest common divisor of 1024 and 100.

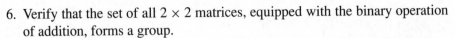

6. Verify that the set of all 2×2 matrices, equipped with the binary operation of addition, forms a group.

7. Verify that the set of all 2×2 matrices, equipped with the binary operation of multiplication, does *not* form a group. What restriction must we make in order for this to be a group?

8. Verify that, in any group,

$$a^2 \cdot b^{-1} = b^{-1} \cdot a^{-2}$$

9. Verify that the set of all polynomials, equipped with the binary operation of multiplication, does *not* form a group.

10. Use the Euclidean algorithm to find the greatest common divisor of each of the following pairs of positive integers.

   (a) 15   80

   (b) 24   92

11. Given an example of integers n and k, with k dividing n, and a finite group of order n that has no subgroup of order k. [*Hint:* The integer k should not be prime.]

CHAPTER 10

Cryptography

10.1 Background on Alan Turing

Alan Mathison Turing was born in 1912 in London, England. He died tragically in 1954 in Wilmslow, Cheshire, England. Today Turing is considered to have been one of the great mathematical minds of the twentieth century. He did not invent cryptography (as we shall see, even Julius Caesar engaged in cryptography). But he ushered cryptography into the modern age. The current vigorous interaction of cryptography with computer science owes its genesis in significant part to the work of Turing. Turing also played a decisive role in many of the key ideas of modern logic. It is arguable that Turing had the decisive ideas for inventing the stored program computer [although it was John von Neumann (1903–1957), another twentieth-century mathematical genius, who together with Herman Goldstine (1913–2004), actually carried out the ideas].

Turing had difficulty fitting in at the British "public schools" which he attended. (Note that a "public school" in Britain is what we in America would call a private school and vice versa.) Young Turing was more interested in pursuing his own thoughts than in applying himself to the dreary school tasks that were designed for average students. At the Sherborne School, Turing had little patience for the tedious math techniques that the teachers taught. Yet he won almost every mathematics prize at the school. He was given poor marks in penmanship, and he struggled with English.

Turing had a passion for science beginning at a very young age. He later said that the book *Natural Wonders Every Child Should Know* had had a seminal influence on him. When he was still quite young, he read Einstein's papers on relativity and he read Arthur Eddington's account of quantum mechanics in the book *The Nature of the Physical World*.

In 1928, at the Sherborne School, Alan Turing became friends with Christopher Morcom. Now he had someone in whom he could confide, and with whom he could share scientific ideas and inquiries. Turing had never derived such intellectual companionship from either his classmates or his rather diffident school teachers. Sadly, Morcom died suddenly in 1930. This event had a shattering effect on the young Alan Turing. The loss of his companion led Turing to consider spiritual matters, and over time this led him to an interest in physics.

It may be mentioned that Turing developed early on an interest in sports. He was a very talented athlete—almost at the Olympic level—and he particularly excelled in running. He maintained an interest in sports throughout his life.

In 1931 Alan Turing entered King's College at Cambridge University. Turing earned a distinguished degree at King's in 1934, followed by a fellowship at King's. In 1936 he won the Smith Prize for his work in probability theory. In particular, Turing was one of the independent discoverers of the Central Limit Theorem.

In 1935 Turing took a course from Max Newman on the foundations of mathematics. Thus his scientific interests took an abrupt shift. The hot ideas of the time were Gödel's incompleteness theorem—which says that virtually any mathematical theory will have true statements in it that cannot be proved—and (what is closely related) David Hilbert's questions about decidability.

In 1936 Alan Turing published his seminal paper "On computable numbers, with an application to the Entscheidungsproblem." Here the *Entscheidungsproblem* is the fundamental question of how to decide—in a manner that can be executed by a machine—when a given mathematical question is provable. In this paper Turing first described his idea for what has now become known as the *Turing machine*. We now take a mathematical detour to talk about Turing machines.

10.2 The Turing Machine

A *Turing machine* is a device for performing effectively computable operations. It consists of a machine through which a bi-infinite paper tape is fed. The tape is divided into an infinite sequence of congruent boxes (Fig. 10.1). Each box has either

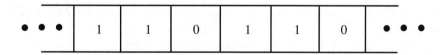

Figure 10.1 A Turing machine.

a numeral 0 or a numeral 1 in it. The Turing machine has finitely many "states" $S_1, S_2, \ldots, S_n$. In any given state of the Turing machine, one of the boxes is being scanned.

After scanning the designated box, the Turing machine does one of three things:

1. Either it erases the numeral 1 that appears in the scanned box and replaces it with a 0, or it erases the numeral 0 that appears in the scanned box and replaces it with a 1, or it leaves the box unchanged.

2. It moves the tape one box (or one unit) to the left or to the right.

3. It goes from its current state S_j into a new state S_k.

It turns out that every logical procedure, every algorithm, every mathematical proof, every computer program can be realized as a Turing machine. The Turing machine is a "universal logical device." The next section contains a simple instance of a Turing machine. In effect, Turing had designed a computer before technology had made it possible to actually build one.

10.2.1 AN EXAMPLE OF A TURING MACHINE

Here is an example of a Turing machine for calculating $x + y$:

State	Old value	New value	Move (L or R)	New state	Explanation
0	1	1	R	0	Pass over x
0	0	1	R	1	Fill gap
1	1	1	R	1	Pass over y
1	0	0	L	2	End of y
2	1	0	L	3	Erase a 1
3	1	0	L	4	Erase another 1
4	1	1	L	4	Back up
4	0	0	R	5	Halt

If you look hard at the logic of this Turing machine, you will see that it thinks of x as a certain number of 1s, and it thinks of y as a certain number of 1s. It scans the x

units, and writes a 1 to the right of these; then it scans y units, and writes a 1 to the right of these. The two blocks of 1s are joined into a single block (by erasing the space in between) and then the two extra 1s are erased. The result is $x + y$. Provide the details of this argument as an exercise.

10.3 More on the Life of Alan Turing

The celebrated logician Alonzo Church (1903–1995) published a paper closely related to Turing's at about the same time. As a result, Church and Turing ended up communicating and sharing ideas. Subsequently, in 1936, Turing went to Princeton for graduate study under Church's direction.

When Turing returned to Cambridge in 1938, he commenced work on actually building a computer. It was designed to be a rather crude, mechanical device, with a great many gears and wheels. In fact Turing had a very specific purpose in mind for his machine.

One of the great mathematical problems of the day (and it is still a hot open problem as of this writing) was to prove the *Riemann hypothesis*. The Riemann hypothesis, posed by Bernhard Riemann in 1859, concerns the location of the zeros of a certain complex function (the celebrated *Riemann zeta function*). An affirmative answer to the Riemann hypothesis would tell us a great deal about the distribution of prime numbers and have profound consequences for number theory and for cryptography.

According to Andrew Hodges (1949–), the Turing biographer,

> Apparently [Turing] had decided that the Riemann Hypothesis was probably false, if only because such great efforts have failed to prove it. Its falsity would mean that the zeta function did take the value zero at some point which was off the special line, in which case this point could be located by brute force, just by calculating enough values of the zeta function.

Turing did his own engineering work, hence he got involved in all the fine details of constructing this machine. He planned on 80 meshing gearwheels with weights attached at specific distances from their centers. The different moments of inertia would contribute different factors to the calculation, and the result would be the location of and an enumeration of the zeros of ζ.

Visits to Turing's apartment would find the guest greeted by heaps of gear wheels and axles and other junk strewn about the place. Although Turing got a good start cutting the gears and getting ready to assemble the machine, more pressing events (such as World War II) interrupted his efforts. His untimely death prevented the completion of the project.

When war broke out in 1939, Turing went to work for the Government Code and Cypher School at Bletchley Park. Turing played a seminal role in breaking German secret codes, and it has been said that his work saved more lives during the war than that of any other person. One of his great achievements during this time was the construction of the *Bombe machine*, a device for cracking all the encoded messages generated by the dreaded German *Enigma machine*. In fact Turing used ideas from abstract logic, together with some earlier contributions of Polish mathematicians, to design the *Bombe*. Turing's important contributions to the war effort were recognized with the award of an O.B.E. (Order of the British Empire) in 1945.

After the war Turing was invited by the National Physical Laboratory in London specifically to design a computer. He wrote a detailed proposal for the Automatic Computing Machine in 1946, and that document is in fact a discursive prospectus for a stored-program computer. The project that Turing proposed turned out to be too grandiose for practical implementation, and it was shelved.

Turing's interests turned to topics outside of mathematics, including neurology and physiology. But he maintained his passion for computers. In 1948 he accepted a position at the University of Manchester. There he became involved in a project, along with F. C. Williams and T. Kilburn, to construct a computing machine.

In 1951 Alan Turing was elected a Fellow of the Royal Society—the highest honor that can be bestowed upon a British scientist. This accolade was largely in recognition of his work on Turing machines.

Turing had a turbulent personal life. In 1952 he was arrested for violation of the British homosexuality statutes. He was convicted, and sentenced to take the drug oestrogen for one year. Turing subsequently rededicated himself to his scientific work, concentrating particularly on spinors and relatively theory. Unfortunately, because of his legal difficulties, Turing lost his security clearance and was labeled something of a "security risk." He had continued working with the cypher school at Bletchley, but his loss of clearance forced that collaboration to end. These events had a profound and saddening effect on Alan Turing.

Turing died in 1954 of potassium cyanide poisoning while conducting electrolysis experiments. The cyanide was found on a half-eaten apple. The police concluded that the death was a suicide, though people close to Turing argue that it was an accident.

10.4 What Is Cryptography?

We use Alan Turing's contributions as a touchstone for our study of cryptography. Cryptography is currently a very hot field, due in part to the availability of high speed digital computers to carry out decryption algorithms, in part to new and exciting connections between cryptography and number theory and logic, and in part to the need for practical coding methods both in industry and in government.

The discussion of cryptography that appears below is inspired by the lovely book [KOB]. We refer the reader to that source for additional ideas and further reading.

As we always do in mathematics, let us begin by introducing some terminology. Cryptography is the study of methods for sending text messages in disguised form in such a manner that only the intended recipient can remove the disguise and read the message. The original message that we wish to send is called the *plaintext* and the disguised message is called the *ciphertext*. We shall always assume that both our plaintext and our ciphertext are written in the standard roman alphabet (that is, the letters A through Z) together perhaps with some additional symbols like "blank space (denoted ⊔)," "question mark (?)," and so forth. The process of translating a plaintext message into a ciphertext message is called *encoding* or *enciphering* or *encrypting*. The process of translating an encoded message back to a plaintext message is called *deciphering* or sometimes *de-encryting*.

For convenience, we usually break up both the plaintext message and the ciphertext message into blocks or units of characters. We call these pieces the *message units*, but we may think of them as "words" (but they are not necessarily English words). Sometimes we will declare in advance that all units are just single letters, or perhaps pairs of letters (these are called *digraphs*) or sometimes triples of letters (called *trigraphs*). Other times we will let the units be of varying sizes—just as the words in any body of text have varying sizes. An *enciphering transformation* is a function that assigns to each plaintext unit a ciphertext unit. The *deciphering transformation* is the inverse mapping that recovers the plaintext unit from the ciphertext unit. Any setup as we have just described is called a *cryptosystem*.

In general it is awkward to mathematically manipulate the letters of the alphabet. We have no notions of addition or multiplication on these letters. So it is convenient to associate to each letter a number. Then we can manipulate the numbers. For instance, it will be convenient to make the assignment

$$A \leftrightarrow 0$$

$$B \leftrightarrow 1$$

$$C \leftrightarrow 2$$

$$\cdots$$

$$X \leftrightarrow 23$$

$$Y \leftrightarrow 24$$

$$Z \leftrightarrow 25$$

Thus if we see the message

$$22 \; 7 \; 0 \; 19 \; 12 \; 4 \; 22 \; 14 \; 17 \; 17 \; 24$$

then we can immediately translate this to

WHATMEWORRY

or

WHAT ME WORRY

Notice that, in cryptography, we generally do not worry about capital and lowercase letters. Everything is uppercase. Second, if we do not have a symbol for "blank space", then messages are awkward to read.

One device of which we will make frequent and consistent use is *modular arithmetic*. Recall that if n and k are an integers then $n \bmod k$ is that unique integer n' between 0 and $k - 1$ inclusive such that $n - n'$ is divisible by k. For example,

$$13 \bmod 5 = 3$$

$$-23 \bmod 7 = 5$$

$$82 \bmod 14 = 12$$

$$10 \bmod 3 = 1$$

How do we calculate these values? Look at the first of these. To determine $13 \bmod 5$, we *divide* 5 into 13: Of course 5 goes into 13 with quotient 2 and remainder 3. It is the *remainder* that we seek. Thus

$$13 \bmod 5 = 3$$

It is similar with the other examples. To determine $82 \bmod 14$, divide 14 into 82. It goes 5 times with remainder 12. Hence

$$82 \bmod 14 = 12$$

It is convenient that modular arithmetic respects the arithmetic operations. For example,

$$8 \times 7 = 56 \qquad \text{and} \qquad 56 \bmod 6 = 2$$

But

$$8 \bmod 6 = 2 \quad \text{and} \quad 7 \bmod 6 = 1 \quad \text{and} \quad 2 \times 1 = 2$$

So it does not matter whether we pass to mod 6 *before* multiplying or *after multiplying*. Either way we obtain the same result 2. Similar properties hold for addition and subtraction. One must be a bit more cautious with division, as we shall see below.

We supply some further examples:

$$[3 \bmod 5] \times [8 \bmod 5] = 24 \bmod 5 = 4$$

$$[7 \bmod 9] + [5 \bmod 9] = 12 \bmod 9 = 3$$

$$[4 \bmod 11] - [9 \bmod 11] = -5 \bmod 11 = 6$$

Now we begin to learn some cryptography by way of examples.

EXAMPLE 10.1
We use the ordinary 26-letter roman alphabet A–Z, with the numbers 0–25 assigned to the letters as indicated above. Let $S = \{0, 1, 2, \ldots, 25\}$. We will consider units consisting of single letters. Thus our cryptosystem will consist of a function $f : S \to S$ which assigns to each unit of plaintext a new unit of ciphertext. In particular, let us consider the specific example

$$f(P) = \begin{cases} P + 5 & \text{if } P < 21 \\ P - 21 & \text{if } P \geq 21 \end{cases}$$

Put in other words,

$$f(P) = P + 5 \bmod 26 \tag{10.1}$$

Next let us use this cryptosystem to encode the message

GOAWAY

or

GO AWAY

The first step is that we transliterate the letters into numbers (because, as noted earlier, numbers are easier to manipulate). Thus GOAWAY becomes 6 14 0 22 0 24.

Now we apply the "shift encryption" [(Eq. 10.1)] to this sequence of numbers. Notice that

$$f(6) = 6 + 5 \bmod 26 = 11 \bmod 26 = 11$$

$$f(14) = 14 + 5 \bmod 26 = 19$$

$$f(0) = 0 + 5 \bmod 26 = 5$$

$$f(22) = 22 + 5 \bmod 26 = 1$$

$$f(0) = 0 + 5 \bmod 26 = 5$$

$$f(24) = 24 + 5 \bmod 26 = 3$$

Thus our ciphertext is 11 19 5 1 5 3. In practice, we may convert this ciphertext back to roman letters using our standard correspondence ($A \leftrightarrow 0$, $B \leftrightarrow 1$, and so on). The result is LTFBFD. Thus the encryption of "GO AWAY" is "LTFBFD." Notice that we have no coding for a blank space, so we ignore it.

This is a very simple example of a cryptosystem. It is said that Julius Caesar used this system with 26 letters and a shift of 3. We call this encryption system a "shift transformation." □

Now let us use this same cryptosystem to encode the word "BRAVO." First, we translate our plaintext word to numbers:

$$1 \quad 17 \quad 0 \quad 21 \quad 14$$

Now we add 5 mod 26 to each numerical entry. The result is

$$6 \quad 22 \quad 5 \quad 0 \quad 19$$

Notice that the fourth entry is 0 because

$$21 + 5 \bmod 26 = 26 \bmod 26 = 0 \bmod 26$$

Thus if we wanted to send the message "BRAVO" in encrypted form, we would send 6 22 5 0 19. We can translate the encrypted message to roman letters as "GWFAT."

Conversely, we *decrypt* a message by subtracting 5 mod 26. Suppose, for instance, that you receive the encrypted message

$$24 \ 12 \ 5 \ 18 \ 15 \ 3 \ 19 \ 25$$

We decrypt applying the function $f^{-1}(Q) = Q - 5 \bmod 26$. The result is

$$19 \ 7 \ 0 \ 13 \ 10 \ 24 \ 14 \ 20$$

This easily translates to

THANKYOU

or

THANK YOU

In a typical, real-life circumstance, you receive an encrypted message and *you do not know the method of encryption*. It is your job to figure out how to decode the message. We call this process *breaking the code*, and the science of codebreaking is called *cryptoanalysis*.

EXAMPLE 10.2
If the codebreaker happens to know that the message he/she has received is encrypted using a shift transformation, then there is a reasonable method to proceed. Imagine that you receive the message

CQNKNJCUNBOXANENA

Looks like nonsense. But the cryptographer has reason to believe that this message has been encoded using a shift transformation on single letters of the 26-letter alphabet. It remains to find the numerical value of the shift.

We use a method called *frequency analysis*. The idea of this technique is that it is known that "E" is the most frequently occurring letter in the English language. Thus we may suppose that the most frequently occuring character in the ciphertext is the encryption of "E" (*not* "E" itself). In fact we see that the character "N" occurs five times in the ciphertext, and that is certainly the most frequently occurring letter. If we hypothesize that "N" is the encryption of "E", then we see that "4" has been translated to "13" in the encryption. Thus the encryption key is $P \mapsto P + 9 \bmod 26$. And therefore the decryption scheme is $P \mapsto P - 9 \bmod 26$. If this putative decryption scheme gives a sensible message, then it is likely the correct choice (as any other decryption scheme will likely give nonsense). Let us try this scheme and see what result it gives. We have

CQNKNJCUNBOXANENA

has numerical realization

2 16 13 10 13 9 2 20 13 1 14 23 0 13 4 13 0

Under our decryption scheme, this translates to

19 7 4 1 4 0 19 11 4 18 5 14 17 4 21 4 17

which has textual realization

THEBEATLESFOREVER

In other words, the secret message is

THE BEATLES FOREVER

The trouble with the shift transformation is that it is just too simpleminded. It is too easy to break. There are variants that make it slightly more sophisticated. For example, suppose that the East Coast and the West Coast branches of National Widget Corporation cook up a system for sending secret messages back and forth. They will use a shift transformation, but *in each week of the year they will use a different shift*. This adds a level of complexity to the process. But the fact remains that, using a frequency analysis, the code can likely be broken in any given week.   □

10.5 Encryption by Way of Affine Transformations

We can add a genuine level of sophistication to the encryption process by adding some new mathematics. Instead of considering a simple shift of the form $P \mapsto P + b$ for some fixed integer b, we instead consider an *affine transformation* of the form $P \mapsto aP + b$. Now we are both multiplying (or dilating) the element P by an integer a and then translating it by b.

10.5.1 DIVISION IN MODULAR ARITHMETIC

There is a subtlety in the application of the affine transformation method that we must consider before we can look at an example. If the encryption scheme is

$P \mapsto Q \equiv aP + b$, then the decryption scheme must be the inverse function. In other words, we solve for P in terms of Q. This just involves elementary algebra, and we find that

$$P = [1/a](Q - b) \bmod 26$$

We see that decryption, in the context of an affine transformation, involves division in arithmetic modulo 26. This is a new idea, and we should look at a couple of simple examples before we proceed with our cryptographic considerations.

We want to consider division modulo 26. Thus if a and b are whole numbers, then we want to calculate b/a and we want the answer to be another whole number modulo 26. This is possible only because we are cancelling multiples of 26, and it will only work when a has no common prime factors with 26. Let us consider some examples.

First let us calculate 4/7 mod 26. What does this mean? We are dividing the whole number 4 by the whole number 7, and this looks like a fraction. But things are a bit different in modular arithmetic. We seek a number k such that

$$\frac{4}{7} \bmod 26 = k$$

or

$$4 = 7 \cdot k \bmod 26$$

or

$$4 - 7 \cdot k \quad \text{is divisible by 26}$$

We simply try different values for k, and we find with $k = 8$ that

$$4 - 7 \cdot 8 = 4 - 56 = -52$$

is indeed divisible by 26. In conclusion,

$$\frac{4}{7} \bmod 26 = 8$$

We see the somewhat surprising conclusion that the fraction 4/7 can be realized as a whole number in arithmetic modulo 26.

Next let us try to calculate $1/4 \mod 26$. This is doomed to fail, because 4 and 26 have the prime factor 2 in common. We seek an integer k such that

$$1 = 4 \cdot k \mod 26$$

or in other words

$$1 - 4k \quad \text{is a multiple of 26}$$

But of course $4k$ will always be even so $1 - 4k$ will always be odd—*it cannot be a multiple of the even number 26.* This division problem cannot be solved.

We conclude this brief discussion with the example $2/9 \mod 26$. We invite the reader to discover that the answer is $6 \mod 26$.

There is in fact a mathematical device for performing division in modular arithmetic. It is the classical Euclidean algorithm. This simple idea is one of the most powerful in all of number theory. It says this: if n and d are integers then d divides into n some whole number q times with some remainder r, and $0 \leq r < d$. In other words,

$$n = d \cdot q + r$$

You have been using this idea all your life when you calculate a long division problem (not using a calculator, of course). We shall see in the next example that the Euclidean algorithm is a device for organizing information so that we can directly perform long division in modular arithmetic.

EXAMPLE 10.3

Let us calculate $1/20$ in arithmetic mod 57. We apply the Euclidean algorithm to 57 and 20. Thus we begin with

$$57 = 2 \cdot 20 + 17$$

We continue by repeatedly applying the Euclidean algorithm to divide the divisor by the remainder:

$$20 = 1 \cdot 17 + 3$$

$$17 = 5 \cdot 3 + 2$$

$$3 = 1 \cdot 2 + 1$$

Now, as previously indicated, we utilize this Euclidean algorithm information to organize our calculations. Begin with the last line to write

$$1 = 3 - 1 \cdot 2$$
$$= 3 - 1 \cdot (17 - 5 \cdot 3)$$
$$= [20 - 17] - 1 \cdot \big([57 - 2 \cdot 20] - 5 \cdot [20 - 17]\big)$$
$$= 20 \cdot 8 + 17 \cdot (-6) - 57$$
$$= 20 \cdot 8 + (57 - 2 \cdot 20) \cdot (-6) - 57$$
$$= 20 \cdot 20 - 7 \cdot 57$$

This calculation tells us that $1 = 20 \cdot 20 \bmod 57$. In other words, $1/20 = 20 \bmod 57$. □

We offer the reader the exercise of calculating $1/25 \bmod 64$ using the Euclidean algorithm.

10.5.2 INSTANCES OF THE AFFINE TRANSFORMATION ENCRYPTION

EXAMPLE 10.4
Let us encrypt the message "GO AWAY" using the affine transformation $P \mapsto 5P + 6 \bmod 26$. As usual,

GO AWAY   has numerical realization 6 14 0 22 0 24

Under the affine transformation, we obtain the new numerical realization

10 24 6 12 6 22

In roman letters, the message has become the ciphertext

KYGMGW

In order to decrypt the message, we must use the *inverse* affine transformation. If $R = 5P + 6 \bmod 26$, then $P = [1/5](R - 6) \bmod 26$. Using modular arithmetic, we see that 10 corresponds to

$$[1/5](10 - 6) = [1/5] \cdot 4 = 6 \bmod 26$$

(because $5 \cdot 6 \bmod 26 = 30 \bmod 26 = 4 \bmod 26$). Likewise 24 corresponds to

$$[1/5](24 - 6) = [1/5] \cdot 18 = 14 \bmod 26$$

(because $5 \cdot 14 \bmod 26 = 70 \bmod 26 = 18 \bmod 26$). We calculate the rest of the correspondences:

$$[1/5](6 - 6) = [1/5] \cdot 0 = 0 \bmod 26$$

(because $5 \cdot 0 \bmod 26 = 0 \bmod 26$). Next,

$$[1/5](12 - 6) = [1/5] \cdot 6 = 22 \bmod 26$$

(because $5 \cdot 22 \bmod 26 = 110 \bmod 26 = 6 \bmod 26$). Again,

$$[1/5](6 - 6) = [1/5] \cdot 0 = 0 \bmod 26$$

And, finally,

$$[1/5](22 - 6) = [1/5] \cdot 16 = 24 \bmod 26$$

(because $5 \cdot 24 \bmod 26 = 16 \bmod 26$).

In sum, we have applied our decryption algorithm to recover the message

$$6 \ \ 14 \ \ 0 \ \ 22 \ \ 0 \ \ 24$$

This transliterates to

GOAWAY

or

GO AWAY   □

In a real-life situation—if we were endeavoring to decrypt a message—we would not know in advance which affine transformation was used for the encoding. We now give an example to illustrate how to deal with such a situation.

EXAMPLE 10.5
We continue to work with the 26-letter roman alphabet. We receive a block of ciphertext and wish to decode it. We notice that the most frequently occurring character in the ciphertext is "M" and the second most frequently occurring character in the ciphertext is "R." It is well known that, in ordinary English, the most commonly occurring letter is "E" and the second most commonly occurring letter is "T." So it is natural to hypothesize that we are dealing with an affine transformation that assigns "E" to "M" and "T" to "R."

This means that we seek an affine transformation $f(P) = aP + b$ such that $f(4) = 12 \bmod 26$ and $f(19) = 17 \bmod 26$. All arithmetic is, as usual, modulo 26. We are led then to the equations

$$12 = a \cdot 4 + b \bmod 26$$

$$17 = a \cdot 19 + b \bmod 26$$

We subtract these two equations to eliminate b and obtain

$$-5 = a \cdot (-15) \bmod 26$$

or

$$a = [-5/(-15)] \bmod 26$$

The solution is $a = 9$. Substituting this value into the first equation gives $b = -24 = 2 \bmod 26$.

Thus our affine encoding transformation is (we hope) $f(P) = 9P + 2$. It is also easy to determine that the inverse (or decoding) transformation is $f^{-1}(Q) = [Q - 2]/9$. □

You Try It: Use the affine decryption scheme in the last example to decode the message "ZMDEMRILMRRMZ."

Next we present an example in which an expanded alphabet is used.

EXAMPLE 10.6
Consider the standard roman alphabet of 26 characters along with the additional characters "blank space" (denoted ⊔), "question mark" (?), "period" (.), and "exclamation point" (!). So now we have 30 characters, and arithmetic will be module 30. As usual, we assign a positive integer to each of our characters.

Thus we have

$$A \leftrightarrow 0$$
$$B \leftrightarrow 1$$
$$C \leftrightarrow 2$$
$$\cdots$$
$$X \leftrightarrow 23$$
$$Y \leftrightarrow 24$$
$$Z \leftrightarrow 25$$
$$\sqcup \leftrightarrow 26$$
$$? \leftrightarrow 27$$
$$. \leftrightarrow 28$$
$$! \leftrightarrow 29$$

Because there are now 30 different characters, we also use 30 different numerical codes—the numbers from 0 to 29.

Imagine that we receive a block of ciphertext, and that we wish to decode it. We notice that the most commonly used characters in the ciphertext are "D" and "!". It is known that the most commonly used characters in ordinary English are "xyz" and "E".[1] If we assume that the ciphertext was encrypted with an affine transformation, then we seek an affine mapping $f(P) = aP + B$ such that $f(\sqcup) = D$ and $f(E) =!$. Thus we are led to $f(26) = 3$ and $f(4) = 29$ and then to the system of equations

$$3 = a \cdot 26 + b \bmod 30$$
$$29 = a \cdot 4 + b \bmod 30$$

As before, we subtract the equations to eliminate b. The result is

$$-26 = 22a \bmod 30$$

---

[1] We formerly said that "E" was the most commonly used letter. But that was before we added the blank space "$\sqcup$" to our alphabet.

This equation is equivalent (dividing by 2) to

$$-13 = 11a \bmod 30$$

Since 11 and 30 have no factors in common, we may easily find the unique solution $a = 7$. Substituting this value in the second equation gives $b = 1$. We conclude that our affine transformation is $f(P) = 7P + 1$.

If the ciphertext we have received is

$$21 \ 7 \ 29 \ 3 \ 14 \ 29 \ 12 \ 14 \ 7 \ 14 \ 19 \ 18 \ 29 \ 24$$

then we can apply $f^{-1}(Q) = [Q - 1]/7$ to obtain the plaintext message

$$20 \ 18 \ 4 \ 26 \ 19 \ 4 \ 23 \ 19 \ 26 \ 18 \ 19 \ 24 \ 11 \ 4 \ 29$$

This transliterates to

USE TEXT STYLE!

A nice feature of this example is that the spaces and the punctuation are built into our system of characters. Hence the translated message is quite clear, and requires no further massaging.   □

10.6 Digraph Transformations

Just to give an indication of how cryptographers think, we shall now consider digraphs. Instead of thinking of our message units as single characters, we will now have units that are *pairs* of characters. Put in other words, the plaintext message is broken up into two-character segments or words. (It should be stressed that these will not, in general, correspond to English words. Certainly words from the English language are generally longer than two letters. Here, when we say "word," we simply mean a unit of information.)

In case the plaintext message has an odd number of characters, then of course we cannot break it up evenly into units of two characters. In this instance we add a "dummy" character like "X" to the end of the message so that an even number of characters will result. Any English message will still be readable if an "X" is tacked on the end.

Let K be the number of elements in our alphabet (in earlier examples, we have seen alphabets with 26 characters and also alphabets with 30 characters). Suppose now that MN is a digraph (that is, an ordered pair of characters from our alphabet). Let x be the numerical equivalent of M and let y be the numerical equivalent of N.

Then we assign to the digraph MN the number $x \cdot K + y$. Roughly speaking, we are now working in base-K arithmetic.

EXAMPLE 10.7
Let us work in the familiar roman alphabet of 26 characters. A common digraph in English is "TH." Notice that the numerical equivalent of "T" is 19 and the numerical equivalent of "H" is 7. According to our scheme, we assign to this digraph the single number $19 \cdot 26 + 7 = 501$.

It is not difficult to see that each positive integer corresponds to a unique digraph. Consider the number 358. Then 26 divides into 358 a total of 13 times with a remainder of 20. We conclude that 358 corresponds to the digraph with numerical equivalents 13 20. This is the digraph "NU."   □

It is straightforward to see that the greatest integer that can arise in this labeling scheme for digraphs is for the digraph $\Omega\Omega$, where Ω is the last character in our alphabet. If the first character is assigned to 0 (as we have done in the past) then the last character is assigned to $K - 1$ (where K is the number of characters in the alphabet). The numerical label is then $(K - 1) \cdot K + (K - 1) = K \cdot K - 1$. So it is safe to say that $K^2 - 1$ is an upper bound for numerical labels in our digraph system.

We conclude, then, that an enciphering transformation is a function that consists of a rearrangement of the integers $\{0, 1, 2, \ldots, K^2 - 1\}$. One of the simplest such transformations is an *affine transformation* on $\{0, 1, 2, \ldots, K^2 - 1\}$. We think of this set of integers as $\mathbb{Z}$ modulo K^2. So the encryption has the form $f(P) = aP + b \bmod K^2$. As usual, the integer a must have no prime factors in common with K^2 (and hence no prime factors in common with K).

EXAMPLE 10.8
We work as usual with the 26-letter roman alphabet. There are then 26×26 digraphs, and these are enumerated by means of the integers $0, 1, 2, \ldots, 26^2 - 1$. In other words, we work in arithmetic modulo 676, where of course $676 = 26^2$. The digraph "ME" has letters "M" corresponding to 12 and "E" corresponding to 4. Thus we assign the digraph number $12 \cdot 26 + 4 = 316 \bmod 676$.

If our affine enciphering transformation is $f(P) = 97 \cdot P + 230$ then the digraph "ME" is encrypted as $97 \cdot 316 + 230 = 462 \bmod 676$.

If instead we consider the digraph "EM" then we assign the integer $4 \cdot 26 + 12 = 116$. And now the encryption is $97 \cdot 116 + 230 = 666 \bmod 676$.   □

EXAMPLE 10.9
Suppose that we want to break a digraphic encryption system that uses an affine transformation. So we need to determine a and b. This will require two pieces of information.

Let us attempt a frequency analysis. From statistical studies, it is known that the some of the most common digraphs are "TH," "HE," and "EA." The most common ones that include the "blank space" character are "E⊔," "S⊔," and "⊔T." If we examine a good-sized block of ciphertext and notice the most commonly occurring digraphs, then we might suppose that those are the encryptions of "TH" or "HE" or "EA." Consider for example the ciphertext (based on the 27-character alphabet consisting of the usual 26 letters of the roman alphabet plus the blank space, and numbered 0 through 26)

$$\text{XIHZYIQHRCZJSDXIDCYIQHPS}$$

We notice that the digraphs "XI," "YI," and "QH" each occur twice in the message. We might suppose that one of these is the encryption of "TH," one is the encryption of "HE," and one is the encryption of "EA" (although, as indicated above, there are other possibilities). Let us attempt to directly solve for the affine transformation that will decript our ciphertext. The affine transformation will have the form $f^{-1}(Q) = a'Q + b'$ and our job is to find a' and b'.

To be specific, let us guess that

$$\text{TH encrypts as YI}$$

$$\text{HE encrypts as XI}$$

This means that we have the numerical correspondences

$$520 \leftrightarrow 656$$

and

$$193 \leftrightarrow 629$$

So we have the algebraic equations

$$520 = a' \cdot 656 + b' \bmod 729$$

$$193 = a' \cdot 629 + b' \bmod 729$$

Subracting the equations as usual (to eliminate b'), we see that

$$327 = a' \cdot 27 \bmod 729$$

Unfortunately this equation does not have a unique solution, because 27 and 729 have prime factors in common (such as 3).

We make another guess. Let us suppose that

TH	encrypts as	QH
HE	encrypts as	YI

This means that we have the numerical correspondences

$$520 \leftrightarrow 439$$

and

$$193 \leftrightarrow 656$$

So we have the algebraic equations

$$520 = a' \cdot 439 + b' \bmod 729$$
$$193 = a' \cdot 656 + b' \bmod 729$$

Subtracting the equations as usual (to eliminate b'), we see that

$$327 = a' \cdot 217 \bmod 729$$

Now 217 and 729 have no prime factors in common, so we may solve for a' uniquely. The answer is $a' = 408$. Substituting into our first equation gives $b' = 13$. So our decryption algorithm is

$$f^{-1}(Q) = 408Q + 13 \tag{10.2}$$

We apply this rule to the ciphertext

$$\text{XIHZYIQHRCZJSDXIDCYIQHPS}$$

For example, the digraph "XI" has numerical equivalent 629. It translates, with decryption rule in Eq. (10.2), to 37. This in turn corresponds to the plaintext digraph "BK." We can already tell we are in trouble, because there is no word in the English language that contains the two letters "BK" in sequence.

It is our job then to try all the other possible correspondences of encrypted digraphs "XI," "YI," and "QH" to the plaintext digraphs. We shall not work them all out here. It turns out that the one that does the trick is

XI is the encryption of TH

and

QH is the encryption of EA

Let us try it and see that it succesfully decrypts our secret message.
The proposed correspondences have numerical interpretation

$$629 \leftrightarrow 520$$

and

$$439 \leftrightarrow 108$$

This leads to the equations

$$520 = a' \cdot 629 + b' \bmod 729$$
$$108 = a' \cdot 439 + b' \bmod 729$$

Subtracting as usual, we obtain

$$412 = a' \cdot 190 \bmod 729$$

Since 190 and 729 have no prime factors in common, we can certainly divide by
190 and solve for a'. We find that $a' = 547$. Substituting into the second equation
gives $b' = 545$. In conclusion, the decrypting transformation is $f^{-1}(Q) = 547Q +
545 \bmod 729$.

Now we can systematically apply this affine transformation to the digraphs in
the ciphertext and recover the original message. Let us begin:

$$XI \rightarrow 629 \xrightarrow{f^{-1}} 520 \rightarrow TH$$
$$HZ \rightarrow 214 \xrightarrow{f^{-1}} 234 \rightarrow IS$$

The calculations continue, and the end result is the original plaintext message

THIS HEART OR THAT HEADX

As you can see, an "X" is affixed to the end to force the message to have an even number of characters (counting blank spaces) so that the digraph method will work.   □

One important point that the last example illustrates is that cryptography will always entail a certain amount of (organized) guesswork.

10.7 RSA Encryption

10.7.1 BASICS AND BACKGROUND

Modern security considerations make it desirable for us to have new types of encryption schemes. It is no longer enough to render a message so that only the intended recipient can read it (and outsiders cannot). In today's complex world, and with the advent of high-speed digital computers, there are new demands on the technology of cryptography. The present section will discuss some of these considerations.

In the old days (beginning even with Julius Caesar), it was enough to have a method for disguising the message that we were sending. For example, imagine that the alphabet is turned into numeric symbols by way of the scheme

$$A \longmapsto 0$$

$$B \longmapsto 1$$

$$C \longmapsto 2$$

and so forth.

Then use an encryption like

$$n \longmapsto n + 3 \bmod 26 \tag{10.3}$$

And now convert these numbers back to roman letters. We have discussed such encryption schemes in the preceding sections.

Today life is more complex. One can imagine that there would be scenarios in which

1. You wish to have a means that a minimum-wage security guard (whom you don't necessarily trust) can check that people entering a facility know a password—but you don't want *him* to know the password.

2. You wish to have a technology that allows anyone to encrypt a message—using a standard, published methodology—but only someone with special additional information can decrypt it.

3. You wish to have a method to be able to convince someone else that you can perform a procedure, or solve a problem, or prove a theorem, without actually revealing the details of the process.

This may all sound rather dreamy, but in fact—thanks to the efforts and ideas of R. Rivest, A. Shamir, and L. Adleman—it is now possible. The so-called RSA encryption scheme is now widely used. For example, the e-mail messages that I receive on my cell phone are encrypted using RSA. Banks, secure industrial sites, high-tech government agences (for example, the National Security Agency), and many other parts of our society routinely use RSA to send messages securely.

In this discussion, we shall describe how RSA encryption works, and we shall encrypt a message using the methodology. We shall describe all the mathematics behind RSA encryption, and shall proved the results necessary to flesh out the theory behind RSA. We shall also describe how to convince someone that you can prove the Riemann hypothesis—without revealing any details of the proof. This is a fascinating idea—something like convincing your mother that you have cleaned your room without letting her have a look at the room. But in fact the idea has profound and far-reaching applications.

10.7.2 PREPARATION FOR RSA

Background Ideas

We now sketch the background ideas for RSA. These are all elementary ideas from basic mathematics. It is remarkable that these are all that are needed to make this profound new idea work.

Computational Complexity

Suppose that you have a deck of N playing cards and you toss them in the air. Now you want to put them back into their standard order. How many "steps" will this take? (We want to answer this question in such a manner that a machine could follow the instructions.)

First we look through all N cards and find the first card in the ordering. Then we look through the remaining $N - 1$ cards and find the second card in the ordering. And so forth.

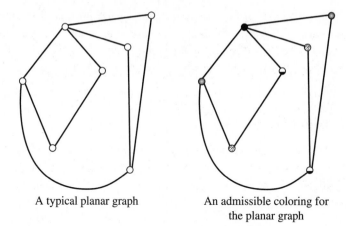

A typical planar graph   An admissible coloring for
                                  the planar graph

Figure 10.2   Coloring of a graph.

So the reordering of the cards takes

$$N + (N - 1) + (N - 2) + \cdots 3 + 2 + 1 = \frac{N(N + 1)}{2}$$

steps. Notice that this answer is a quadratic polynomial in N. Thus we say that the problem can be solved in polynomial time.

We have heard a rumor that the four-color theorem is true. So we have a graph with N vertices and we wish to color each vertex, using either red, yellow, blue, or green. See Fig. 10.2, where we have used shading to suggest the coloring. The only rule is that two adjacent vertices (that is, vertices that are connected by a segment) cannot be the same color.

Of course the number of possible colorings is the number of functions from the set with N objects to the set with 4 objects. That is 4^N. The machine, being as dumb as it is, will simply try all the possible colorings until it finds one that works. Thus we see that the number of steps is now an exponential function of N.

We call this an exponential time problem.

Another interesting exponential time problem is that of scheduling planes for an airline. If you have n cities and k planes and you take into account different populations, different demands, crew availability, fuel availability, and other factors, then it is easy to convince yourself that this is a problem of exponential complexity. The theory of linear programming can be used to reduce many problems of this kind to polynomial complexity. Linear programming is routinely used by the airlines for this purpose.

Certainly one of the most famous exponential time problems is the "traveling salesman problem." The issue here is that a certain salesman wants to visit each of n cities precisely once. What is his most efficient path? It is not difficult to discern that there are exponentially many possible paths, and no evident strategy for picking one in any efficient manner.

Modular Arithmetic

This is a familiar idea, and we have already alluded to it earlier. The "right" way to define the idea is with cosets, but we shall content ourselves here with a more informal definition.

When we write $n \bmod k$ we mean simply the remainder when n is divided by k. Thus

$$25 = 1 \bmod 3$$

$$15 = 3 \bmod 4$$

$$-13 = -3 \bmod 5 = 2 \bmod 5$$

It is an important fact—which again is most clearly seen using the theory of cosets—that modular arithmetic respects sums and products. That is,

$$a + b \bmod n = a \bmod n + b \bmod n \quad \text{and} \quad a \cdot b \bmod n = (a \bmod n) \cdot (b \bmod n)$$

We shall use these facts in a decisive manner below.

Fermat's Theorem

Let a and b be two (positive) integers. We say that a and b are relatively prime if they have no common prime factors. For example,

$$72 = 2^3 \cdot 3^2$$

$$175 = 5^2 \cdot 7$$

hence 72 and 175 are relatively prime.

If n is an integer, let $\mathcal{P}(n)$ be the set of integers less than n that are relatively prime to it. Let $\varphi(n)$ be the number of elements in $\mathcal{P}(n)$.

Theorem 10.1 *If n is a positive integer and k is relatively prime to n then*

$$k^{\varphi(n)} = 1 \bmod n$$

Proof: The proof of this result is easy. For the collection $\mathcal{P}(n)$ of numbers relatively prime to n forms a group under multiplication. That is, if a is relatively prime to n and b is relatively prime to n then logic dictates that $a \cdot b$ is relatively prime to n. Now it is a fundamental fact—we cannot prove it here, but see [BMS]—that if a group has m elements and g is an element of the group then g^m is the group identity. Thus any element of the group, raised to the power $\varphi(n)$ (the number of elements in the group) will equal 1 modulo n.   □

For later use, it is worth noting that if p, q are prime numbers and $n = p \cdot q$ then $\varphi(n) = (p - 1) \cdot (q - 1)$.

The reason is that the only numbers less than or equal to n that are not relatively prime to n are $p, 2p, 3p, \ldots q \cdot p$ and $q, 2q, 3q, \cdots (p - 1)q$.

There are q numbers in the first list and $p - 1$ numbers in the second list. The set $\mathcal{P}(n)$ of numbers relatively prime to n is the complement of these two lists, and it therefore has

$$pq - q - (p - 1) = pq - q - p + 1 = (p - 1) \cdot (q - 1) \equiv \varphi(n)$$

elements.

Relatively Prime Integers

Two integers a and b are relatively prime if they have no prime factors in common. As noted above, for example, 72 and 175 are relatively prime.

It is a fundamental fact of elementary number theory that if a, b are relatively prime then we can find other integers x and y such that

$$xa + yb = 1 \tag{10.4}$$

For example, we have noted that $a = 72$ and $b = 175$ are relatively prime. The corresponding integers x, y are $x = -17$ and $y = 7$. Thus

$$(-17) \cdot 72 + 7 \cdot 175 = 1$$

One can prove this result using Fermat's theorem above. For, since b is relatively prime to a, thus

$$b^{\varphi(a)} = 1 \bmod a$$

But this just says that

$$b^{\varphi(a)} - 1 = k \cdot a$$

for some integer k. Unraveling this gives Eq. (10.4).

In practice, one finds x and y using the Euclidean algorithm (otherwise known as long division).

In the example of 72, 175, one calculates:

$$175 = 2 \cdot 72 + 31$$
$$72 = 2 \cdot 31 + 10$$
$$31 = 3 \cdot 10 + 1$$

You know you are finished when the remainder is 1.

For now we have

$$1 = 31 - 3 \cdot 10$$
$$= 31 - 3 \cdot (72 - 2 \cdot 31)$$
$$= 7 \cdot 31 - 3 \cdot 72$$
$$= 7 \cdot (175 - 2 \cdot 72) - 3 \cdot 72$$
$$= 7 \cdot 175 - 17 \cdot 72$$

That is the decomposition we seek.

10.7.3 THE RSA SYSTEM ENUNCIATED

Description of RSA

Now we can quickly and efficiently describe how to implement the RSA encryption system, and we can explain how it works.

Imagine that George W. Bush has an important message that he wishes to send to Donald Rumsfeld. Of course Rumsfeld is a highly placed man of many responsibilities, and you can imagine that Bush's message is quite secret. So he wants to encode the message:

> Your time is up. Hasta la vista, baby.

So Bush goes to the library and finds the RSA encryption book. This is a readily available book that anyone can access. It is not secret. A typical page in the book reads like this:

Name	Value of n	Value of e
Puck, Wolfgang	4431 ... 7765	8894 ... 4453
Rehnquist, William	6668 ... 2345	1234 ... 9876
Riddle, Nelson	7586 ... 2390	4637 ... 4389
Rin Tin-Tin	5355 ... 5353	5465 ... 7648
Rogers, Roy	7859 ... 4359	3058 ... 2934
Roosevelt, Theodore	7835 ... 2523	7893 ... 4232
Rotten, Johnny	3955 ... 4343	4488 ... 9922
Roy, Rob	3796 ... 5441	2219 ... 3319
Rumsfeld, Donald	1117 ... 8854	9266 ... 2388
Russert, Tim	6464 ... 4646	3223 ... 3232
Schwarzenegger, Arnold	6894 ... 3242	7525 ... 2314
Simpson, Orenthal James	6678 ... 2234	4856 ... 2223

What does this information mean? Of course we know, thanks to Euclid, that there are infinitely many primes. So we can find prime numbers with as many digits as we wish. Each number n in the RSA encryption book is the product of two 75-digit primes p and q: Thus $n = p \cdot q$. Each number e is chosen to be a number with at least 100 digits that is relatively prime to $\varphi(n) = (p - 1) \cdot (q - 1)$. Of course we do not publish the prime factorization of the number n; we also do not publish $\varphi(n)$. All that we publish is n and e for each individual.

Now an important point to understand is that Bush does not need to understand any mathematics or any of the theory of RSA encryption in order to encode his message. (Well, it would be nice if he understood modular arithmetic. But he is, after all, the President of the United States.) All he does is this:

1. First he breaks the message into units of five letters. We call these "words", even though they may not be English language words.

   For the message from Bush to Rumsefeld, the "words" would be

   YOURT IMEIS UPHAS TALAV ISTAB ABY

2. He transliterates each "word" into a sequence of numerical digits, using our usual scheme of translation.

3. Then he encodes each transliterated word w with the rule

$$w \longmapsto w^e \bmod n$$

 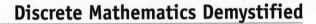

Bush will send to Rumsfeld this sequence of encrypted words. That is all there is to it.

The real question now is:

What does it take to decrypt the encoded message?
How can Rumsfeld read the message?

This is where some mathematics comes into the picture. We must use Fermat's theorem, and we must use our ideas about relatively prime integers. But the short answer to the question is this. If $\widetilde{w}$ is a word encrypted according to the simple scheme described above, then we decrypt it with this algorithm:

We find integers x and y so that $xe + y\varphi(n) = 1$ and then we calculate

$$\widetilde{w}^x \bmod n$$

That will give the decrypted word w with which we began. [We shall provide the mathematical details of this assertion in the next section.] Since w has only five characters, and n has 150 characters, we know that $w \bmod n = w$—so there is no ambiguity arising from modular arithmetic. We can translate w back into roman characters, and we recover our message.

Now here is the most important point in our development thus far:

In order to encrypt a message, we need only look up n and e in the public record RSA encryption book. But, in order to decrypt the message, we must know x. Calculating x necessitates knowing $\varphi(n)$, and that necessitates knowing the prime factorization of n.

It is a theorem that calculating the prime factorization of an integer with k digits is a problem of exponential complexity in k. For an integer with 150 digits, using a reasonably fast computer, it would take several years to find the prime factorization.[2]

10.7.4 THE RSA ENCRYPTION SYSTEM EXPLICATED

Explanation of RSA

In fact, with all the preliminary setup that we have in place, it is a simple matter to explain the RSA encryption system.

---

[2]One might note that—if your message is just five letters long—then there are only 26^5 possible encryptions. You could decrypt the message by trial and error. Because of considerations like this, we often find it convenient to append a 50-digit random integer to the message. This technique is discussed in detail near the end of this discussion.

For suppose that we selected an $n = p \cdot q$ and an e relatively prime to $\varphi(n) = (p-1) \cdot (q-1)$ corresponding to a particular person listed in the RSA encryption book. If we are the certified decrypter, then we know the prime factorization of n—that is, we know that $n = p \cdot q$ for p and q prime.

We therefore know $\varphi(n) = p \cdot q - p - q + 1 = (p-1) \cdot (q-1)$ and so we can calculate the x and the y in the identity $xe + y\varphi(n) = 1$. Once we know x, then we know everything. For

$$\widetilde{w}^x \bmod n = [w^e]^x \bmod n$$

$$= w^{ex} \bmod n$$

$$= w^{1-y\varphi(n)} \bmod n$$

$$= w \cdot [w^{\varphi(n)}]^{-y} \bmod n$$

$$= w \cdot 1^{-y} \bmod n$$

$$= w \bmod n$$

$$= w$$

since w is certainly relatively prime to n.

This shows how we recover the original word w from the encrypted word $\widetilde{w} = w^e \bmod n$.

10.7.5 ZERO KNOWLEDGE PROOFS

How to Keep a Secret

We shall now give a quick and dirty description of how to convince someone that you can prove **Proposition A** without revealing any details of the proof of **Proposition A**. The idea comes across most clearly if we deal again with colorings of graphs.

So suppose once again that we are given a graph. See Fig. 10.3. We are the prover, and there is a remotely located verifier.

It is our job to convince the verifier that we know how to four-color this graph. But we do not want the verifier to actually know how to color the graph. We only want him to be convinced that we know how to do it.

We begin by transmitting the adjacency matrix to the verifier. This data is exhibited in Table 10.1.

This transmission is straightforward, and need not be encoded. We simply tell the verifier: "In position $(1, 1)$ of the matrix there is an x;" "In position $(1, 2)$ of the matrix there is an x;" "In position $(1, 5)$ of the matrix there is no x." And so forth.

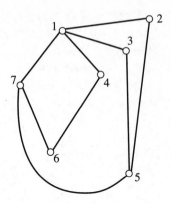

Figure 10.3 Zero knowledge coloring of a graph.

Now we number the colors 1, 2, 3, and 4. Here

$$1 \leftrightarrow \text{blue}$$

$$2 \leftrightarrow \text{red}$$

$$3 \leftrightarrow \text{yellow}$$

$$4 \leftrightarrow \text{green}$$

Finally, imagine that our coloring of the graph is as in Fig. 10.4. Of course we have used shading to suggest the coloring. We wish to communicate to the verifier that we have a valid coloring—in such a manner that he can check it, but he cannot learn any of the details of the coloring.

As we see from Figure 10.4, the coloring is encoded as

$$12 \quad 23 \quad 34 \quad 41 \quad 51 \quad 64 \quad 73$$

Table 10.1

	1	2	3	4	5	6	7
1	x	x	x	x			x
2	x	x			x		
3	x		x		x		
4	x			x		x	
5		x	x		x		x
6			x		x	x	x
7	x				x	x	x

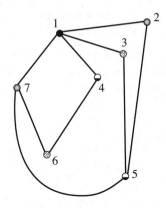

Figure 10.4   Coding of the coloring.

This is read as "node 1 is colored with color 2 (red)," "node 2 is colored with color 3 (yellow)," and so on. We will transmit these pairs of digits, suitably encoded, to the verifier.

   The trouble is that the verifier already knows that there are only seven nodes in the graph, and only four colors, so he could (with a little effort) figure out what color has been assigned to what node just with a little trial and error—even though the information has been encoded. So this will not do.

   Thus, instead of encoding and sending 12 and 23 and 34 and so forth, instead the prover encodes and sends

$$12\, r_1$$

$$23\, r_2$$

$$34\, r_3$$

and so on

where r_1, r_2, r_3 are 50-digit random integers.

   More precisely, the step-by-step scenario is this:

1. The prover sends the entire coloring to the verifier, in encoded form as indicated above.

2. The verifier stares at his adjacency matrix. He notices that, for example, vertices 4 and 6 are adjacent. And he inquires specifically about those two vertices.

3. The prover sends the verifier the colorings for those two particular vertices (with a 50-digit random integer attached to each, just as before).

4. The verifier encrypts the information for those two vertices—using the pre-agreed-upon public key encryption system. He then checks that those two vertex colorings match colorings in the full coloring of the graph that was sent in Step 1.

Now the verifier has checked that one pair of adjacent vertices is suitably colored (that is, with different colors). If he wants to perform further verifications, then the preceding steps are repeated. Except that first the prover assigns numbers to the four colors red, yellow, blue, and green in some new random fashion. And he chooses an entirely new set of random 50-digit integers. Then he sends the entire colored graph, gets a query from the verifier, and so forth.

If there are n nodes to be colored then there are $n(n-1)/2$ possible pairs of nodes. The probability that the prover *lied* about the coloring and that the verifier—in asking for the coloring of a particular pair of nodes—failed to catch the lie is $1/[n(n-1)]$. If the entire process is iterated again, then the probability that the verifier failed to catch the lie is $1/[n^2(n-1)^2]$. And so forth. Thus each successive verification increases the likelihood that the verifier may be certain of his check.

The point of this procedure is that the verifier can check that any pair of adjacent vertices is colored correctly, that no two adjacent vertices are colored the same, but he cannot amalgamate the information and produce the entire coloring of the graph.

Of course the example we have presented is for graph coloring, just because that is simple to describe. But any proof whatsoever can be translated into binary code and then rendered as a statement about the coloring of some graph. So in fact the example we have given is perfectly general.

Concluding Remarks

The RSA encryption scheme is one of the great ideas of modern coding theory. It is being developed and enhanced even as we speak. There are versions for multiple verifiers, for dishonest provers, and many other variants. The history of RSA is a remarkable one. There was a talk at the 1986 International Congress of Mathematicians in Berkeley about the method. After that, the government attempted to co-opt the method, retract all the preprints, and suppress the information. Interestingly, it was the National Security Agency (the branch of the government in Washington that specializes in cryptography) that stepped in and prevented the government intervention. Now RSA is in the public domain, and anyone can use it. It is a powerful tool.

Exercises

1. Use the shift encryption system given by $P \mapsto P - 3$ to encrypt the message

   BYE BYE, BIRDIE

2. Use the shift *decription* scheme $P \mapsto P - 12$ to decrypt the code EAXAZSNMNK.

3. Use a frequency analysis on the ciphertext ZRRGZRURER to determine the shift encryption scheme. Then decrypt the message.

4. Use the affine encryption system given by $P \mapsto 3P + 11$ to encrypt the message

   HELLO MY HONEY

5. Use the affine *decryption* scheme $P \mapsto [P - 3]/7$ to decrypt the code RDQYPHZYDQYP.

6. Break the message

   THIS WAS NOT THE END

   up into two-character digraphs. Now tranlate each digraph into a pair of numbers, and then encrypt each digraph according to the rule $P \mapsto 3P + 7$. Now translate back to a new encrypted word expressed with roman characters.

7. Consider the message

   NOW IS THE TIME

   Transliterate this to a list of numerals, one character at a time, in the usual way. Now apply the encryption algorithm

   $$P \mapsto 5P^2 + P \bmod 26$$

   What ciphertext results?

CHAPTER 11

Boolean Algebra

11.1 Description of Boolean Algebra

11.1.1 A SYSTEM OF ENCODING INFORMATION

Boolean algebra, named after George Boole (1815–1864), is a formal system for encoding certain relationships that occur in many different logical systems. As an example, consider the algebra of sets, equipped with the operations $\cap$, $\cup$, and $^c(\ )$. These are "intersection," "union," and "complementation." Propositional logic has three analogous operations: $\wedge$, $\vee$, and $\sim$. If we think of these as corresponding,

$$\cap \longleftrightarrow \wedge$$

$$\cup \longleftrightarrow \vee$$

$$^c(\ ) \longleftrightarrow \sim$$

then we find that the two logical systems have very similar formulas:

(i) $[^c(S \cup T) = {}^cS \cap {}^cT] \longleftrightarrow$

$$[\sim (A \vee B) \Longleftrightarrow \sim A \wedge \sim B]$$

(ii) $[^c(S \cap T) = {}^cS \cup {}^cT] \longleftrightarrow$

$$[\sim (A \wedge B) \Longleftrightarrow \sim A \vee \sim B]$$

(iii) $[S \cup (T \cap U) = (S \cup T) \cap (S \cup U)] \longleftrightarrow$

$$[A \vee (B \wedge C) \Longleftrightarrow (A \vee B) \wedge (A \vee C)]$$

(iv) $[S \cap (T \cup U) = (S \cap T) \cup (S \cap U)] \longleftrightarrow$

$$[A \wedge (B \vee C) \Longleftrightarrow (A \wedge B) \vee (A \wedge C)]$$

Other logical systems, such as the theory of gates in computer logic, or the theory of digital circuits in the basic theory of electricity, satisfy analogous properties. Boolean algebra abstracts and unifies all of these ideas into a single algebraic system.

The *Stone's representation theorem* [JOH] states that every boolean algebra can be realized as the algebra of closed-and-open sets on a compact, zero-dimensional topological space. This unifies several different areas of mathematics and has proved to be a powerful point of view.

11.2 Axioms of Boolean Algebra

11.2.1 BOOLEAN ALGEBRA PRIMITIVES

Boolean algebra contains these primitive elements: a collection, or set, S of objects. At a minimum, S will contain the particular elements 0 and 1. Boolean algebra also contains three operations (two binary and one unary): $+$, $\times$, and $^-$. Boolean algebra uses the equal sign $=$ and parentheses $(,)$ in the customary manner. In boolean algebra, just as in fuzzy set theory, we think of the overbar as denoting set complementation (although it could have other specific meanings in particular contexts).

11.2.2 AXIOMATIC THEORY OF BOOLEAN ALGEBRA

The axioms for boolean algebra, using elements $a, b, c \in S$, are these:

1. $a \times 1 = a$
2. $a + 0 = a$

3. $a \times b = b \times a$

4. $a + b = b + a$

5. $a \times (b + c) = (a \times b) + (a \times c)$

6. $a + (b \times c) = (a + b) \times (a + c)$

7. $a \times a = 1$

8. $a \times \bar{a} = 0$

9. $a + \bar{a} = 1$

Some of these axioms (1, 2, 3, 4, 5) have a familiar form that we have seen in ordinary arithmetic. The other four (6, 7, 8, 9) are not familiar, and may be counterintuitive. In fact Axioms (6, 7, 8, 9) are *false* in ordinary arithmetic. So Peano's arithmetic is *not* a model for boolean algebra. But the algebra of sets *is* a model for boolean algebra. If we let S be the collection of all sets of real numbers, and if we interpret $+$ as union ($\cup$), $\times$ as intersection ($\cap$), $^-$ as set-theoretic complementation ($^c(\ )$), 0 as the empty set ($\emptyset$), and 1 as the universal set (in this model, the real numbers $\mathbb{R}$), then in fact all nine axioms now hold. As an example,

$$S \cap {}^c S = \emptyset$$

is the correct interpretation of Axiom (8), and it is now true. Also

$$S \cup (T \cap U) = (S \cup T) \cap (S \cup U)$$

is the correct interpretation of Axiom (6), and it is now true. Similar statements may be made about the interpretation of the axioms of boolean algebra in the propositional calculus.

11.2.3 BOOLEAN ALGEBRA INTERPRETATIONS

Obversely, we can also give the boolean algebra interpretations of the four sets of equivalent statements that we gave in Sec. 11.1.1. These are

(i) $\overline{(a + b)} = \bar{a} \times \bar{b}$

(ii) $\overline{(a \times b)} = \bar{a} + \bar{b}$

(iii) $a + (b \times c) = (a \times b) + (a \times c)$

(iv) $a \times (b + c) = (a \times b) + (a \times c)$

11.3 Theorems in Boolean Algebra

11.3.1 PROPERTIES OF BOOLEAN ALGEBRA

One of the remarkable features of boolean algebra is that it has a very small set of axioms, yet many additional desirable properties are readily derived. Some of these properties are:

1. $a \times a = a$
2. $a + a = a$
3. $a \times 0 = 0$
4. $a + 1 = 1$
5. $0 \neq 1$
6. 0 is unique and 1 is unique
7. $\bar{\bar{a}} = a$
8. $a + (b + c) = (a + b) + c$
9. $a \times (b \times c) = (a \times b) \times c$
10. $\overline{(a \times b)} = \bar{a} + \bar{b}$
11. $\overline{(a + b)} = \bar{a} \times \bar{b}$
12. $\bar{0} = 1$
13. $\bar{1} = 0$
14. $a \times (a + b) = a$
15. $a + (a \times b) = a$
16. $\bar{a} \times (a \times b) = 0$
17. $\bar{a} + (a + b) = 1$
18. $a \times (\bar{a} \times \bar{b}) = 0$
19. $a + (\bar{a} + \bar{b}) = 1$
20. If $a \times c = b \times c$ and $a + c = b + c$, then $a = b$

11.3.2 A SAMPLE PROOF

As an illustration of the ideas, we now provide a proof of formula **2**. The proof consists of a sequence of statements, each justified by one of the nine axioms of boolean algebra or by the rules of logic. At each step, we cite the relevant axiom or rule.

1. [**Axiom (2)**] $a + 0 = a$
2. [**Axiom (4)**] $a = a + 0$

3. **[Axiom (7)]** $a \times \overline{a} = 0$
4. **[Axiom (3)]** $0 = a \times \overline{a}$
5. **[Steps 2 and 4]** $a = a + (a \times \overline{a})$
6. **[Axiom (6)]** $a = (a + a) \times (a + \overline{a})$
7. **[Axiom (8)]** $a = (a + a) \times 1$
8. **[Axiom (1) + Step 4 + Step 6]** $a = a + a$
9. **[Symmetry of equality]** $a + a = a$

Some of the elementary theorems of boolean algebra that we have cited require rather elaborate justifications. For instance, it is rather difficult to prove the associativity of addition. We refer the reader to [NIS] for the details.

11.4 Illustration of the Use of Boolean Logic

We now present an example adapted from one that is presented in [NIS, p. 41]. Imagine an alarm system for the monitoring of hospital patients. There are four factors (inputs/outputs) that contribute to the triggering of the alarm:

Inputs/Outputs	Meaning
a	Patient's temperature is in the range 36–40°C.
b	Patient's systolic blood pressure is outside the range 80–160 mm
c	Patient's pulse rate is outside the range 60–120 beats per minute
o	Raising the alarm is necessary

Good sense dictates that we would want the alarm to sound if any of the following situations obtains:

- The patient's temperature is outside the acceptable range, the blood pressure is outside the acceptable range, and the pulse rate is outside the acceptable range.

- The patient's temperature is in the acceptable range, but the pulse rate is outside the acceptable range.

- The patient's temperature is in the acceptable range, but the systolic pressure is outside the acceptable range.

- The patient's temperature is in the acceptable range, but both the systolic pressure and the pulse rate are outside the acceptable range.

11.4.1 BOOLEAN ALGEBRA ANALYSIS

Using boolean algebra, these four conditions can be encoded as

- $\bar{a} \times b \times c$
- $a \times \bar{b} \times c$
- $a \times b \times \bar{c}$
- $a \times b \times c$

Notice that, since we know that multiplication is associative, we have omitted using parentheses to group these binary operations. No ambiguity results.

The aggregate of all the situations in which we want the alarm to sound can be represented by the equation

$$o = [\bar{a} \times b \times c] + [a \times \bar{b} \times c] + [a \times b \times \bar{c}] + [a \times b \times c] \qquad (11.1)$$

Now we can use the laws of boolean algebra to simplify the right-hand side:

$$[\bar{a} \times b \times c] + [a \times \bar{b} \times c] + [a \times b \times \bar{c}] + [a \times b \times c]$$

$$= [\bar{a} \times b \times c] + [a \times \bar{b} \times c] + [a \times b \times \bar{c}]$$

$$+([a \times b \times c] + [a \times b \times c] + [a \times b \times c])$$

$$= ([\bar{a} \times b \times c] + [a \times b \times c])$$

$$+([a \times \bar{b} \times c] + [a \times b \times c])$$

$$+([a \times b \times \bar{c}] + [a \times b \times c])$$

$$= ([\bar{a} + a] \times [b \times c])$$

$$+([\bar{b} + b] \times [a \times c])$$

$$+([\bar{c} + c] \times [a \times b])$$

$$= (1 \times [b \times c]) + (1 \times [a \times c]) + (1 \times [a \times b])$$

$$= [b \times c] + [a \times c] + [a \times b]$$

In summary, we have reduced our protocol for the sounding of the medical alarm to:

$$o = [b \times c] + [a \times c] + [a \times b] \tag{11.2}$$

The circuit that we put in place can be designed after Eq. (11.2) rather than after the much more complicated Eq. (11.1).

Exercises

1. As much as possible, simplify the boolean expression

$$[\bar{a} \times \bar{b} \times c] + [a \times \bar{b} \times \bar{c}] + [\bar{a} \times b \times \bar{c}] + [\bar{a} \times \bar{b} \times \bar{c}]$$

2. Prove the boolean identity

$$a \times (a + b) = a$$

3. Prove the boolean identity

$$a \times (\bar{a} \times \bar{b}) = 0$$

4. Prove the boolean identity

$$\overline{(a + b)} = \bar{a} \times \bar{b}$$

5. Prove the boolean identity

$$\overline{(a \times b)} = \bar{a} + \bar{b}$$

6. Prove the boolean identity

$$a + (a \times b) = a$$

7. Prove that if $a \times c = b \times c$ and $a + c = b + c$, then $a = b$.

CHAPTER 12

Sequences

12.1 Introductory Remarks

Many physical and mathematical quantities are best understood by using approximations. Sometimes approximating values are given as limits. For instance, suppose that we want to paint the (infinitely many) squares shown in Fig. 12.1. Suppose also that 1 gallon of paint covers 500 square feet. How much paint will we need?

We see that the first square has interior area 900 square feet. The first *two* squares have area $900 + 90 = 990$ square feet. The first *three* squares have area $900 + 90 + 9 = 999$ square feet. The first *four* squares have area $900 + 90 + 9 + 0.9 = 999.9$ square feet. The pattern is clear. We have a list of approximations, each given by a sum of finitely many numbers. The approximations seem to tend to 1000. Therefore it seems that we will need enough paint to cover 1000 square feet, or 2 gallons of paint ([we are of course conveniently ignoring the fact that the (100^{100})th square will be too small to hold even one molecule of paint)].

The purpose of this chapter is to turn the above intuitive discussion into careful mathematics. A list of approximating values will be called a "sequence." We want especially to study the situation when the sequence consists of "partial sums" (as in

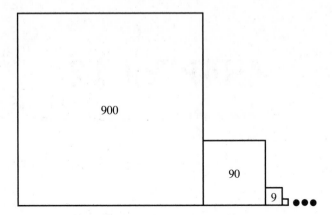

Figure 12.1   Calculating the area of a floor.

the paint example)—this is called a "series." Thus the process we will be performing corresponds, as in the example of the squares, to summing infinitely many numbers.

12.2 Infinite Sequences of Real Numbers

A *sequence* is an ordered list of numbers. It is most common to write a sequence as

$$a_1, a_2, a_3, \ldots$$

or sometimes as

$$\{a_j\}_{j=1}^{\infty}$$

EXAMPLE 12.1
Discuss the sequence $2, 1, 4, 3, 6, 5, \ldots$.

Solution: We see that

$$a_1 = 2, \quad a_2 = 1, \quad a_3 = 4, \quad a_4 = 3, \quad a_5 = 6, \quad a_6 = 5, \ldots$$

In fact the *rule* that generates the sequence assigns to each positive odd integer the next integer and to each positive even integer the preceding integer. Check this assertion: $f(1) = 2, f(2) = 1, f(3) = 4, f(4) = 3, f(5) = 6, f(6) = 5$, and so on.   □

Remark 12.1 It would be a mistake to think that every sequence is given by a "rule." Far from it. But many sequences do come from rules, and it is always interesting to determine what that rule is.

EXAMPLE 12.2
How does the sequence

$$1, 2, 3, 4, \ldots$$

differ from the sequence in Example 12.1?

Solution: This new sequence has the same *values* as the sequence in Example 12.1. But they occur in a different order. Since a sequence is by definition an *ordered list*, we conclude that this is a different sequence from that in Example 12.1.   □

Insight: Occasionally it will prove useful to begin a sequence with an index different from 1. An example is

$$\{3j - 5\}_{j=4}^{\infty}$$

This denotes the sequence 7, 10, 13, 16, ...

Generally, the main reason for studying a sequence is to understand whether or not it "tends to some limit." Before giving a precise definition of limit, let us apply our intuition to some examples.

EXAMPLE 12.3
Discuss whether the sequence

$$a_j = 2^{-j}$$

which can also be written as

$$\{2^{-j}\}_{j=1}^{\infty}$$

tends to a limiting value.

Solution: If we write the sequence out as

$$\frac{1}{2}, \frac{1}{4}, \frac{1}{8}, \ldots$$

Figure 12.2   The limit of a sequence.

then we see that the terms become (and remain) arbitrarily close to zero. It seems plausible to say that the sequence tends to zero. This intuitive notion is displayed in Fig. 12.2.   □

EXAMPLE 12.4
Does the sequence $a_j = (-1)^j$ tend to a limit?

Solution: We may write this sequence out as

$$-1, 1, -1, 1, \ldots$$

The sequence does not seem to tend to any limit: half of the time the value is 1 and the other half of the time the value is -1. The sequence does not *become* and *remain* close to a single value. Therefore we say that it has no limit. Refer to Fig. 12.3.   □

EXAMPLE 12.5
Does the sequence $a_j = j^3$ tend to a limit?

Solution: This sequence may be written out as

$$1, 8, 27, 64, \ldots$$

The sequence takes values which are larger and larger, without any bound. The sequence tends to no limit. Refer to Fig. 12.4.   □

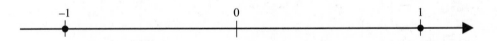

Figure 12.3   A sequence with no limit.

Figure 12.4   A sequence that grows without bound and hence has no limit.

EXAMPLE 12.6

A quantity of radioactive material decays. At the beginning of each week there is half as much as there was the previous week. The initial quantity is 5 grams. Use sequence notation to express the amount of material at the beginning of the jth week.

Solution: The amount of radioactive material at the beginning of the second week is 5/2 (half as much as the initial amount at the beginning of the first week). The amount at the beginning of the third week is 5/4. The amount at the beginning of the fourth week is 5/8. And so forth.

   As a result, according to the description, the amount of material at the start of the jth week is

$$a_j = 5 \cdot \left(\frac{1}{2}\right)^{j-1}$$

The sequence exhibits in an elegant way the process of radioactive decay: the first several values are

$$5, \frac{5}{2}, \frac{5}{4}, \frac{5}{8}$$

It is easy to see intuitively, or with a calculator, that the amount of radioactive material tends to 0 as time tends to ∞.   □

EXAMPLE 12.7

Discuss convergence for the sequence 1 , 1/2 , 1/3 , 1/4 , ...

Solution: Each term is smaller than the last. All the terms are positive. Most importantly, the terms are getting arbitrarily close to zero: the hundredth term is within 1/100 of 0, the thousandth term is within 1/1000 of 0, the millionth term is within 1/1000000 of 0. We conclude that the sequence converges, and its limit is 0.   □

EXAMPLE 12.8

Let $a_j = c$ for every j. What is the limit of this constant sequence?

Solution: The terms of this sequence are all the same. They do not deviate. They stay at c—never above c and never below c. They are arbitrarily close to c because in fact they all *equal* c. We conclude that the sequence converges, and its limit is c. $\square$

EXAMPLE 12.9
Find the limit of the sequence

$$a_j = \frac{2j}{j+1}$$

Solution: It is helpful to use long division and write

$$a_j = 2 + \frac{-2}{j+1}$$

Now the numerator in the fraction on the far right is plainly the constant value -2, while the denominator increases without bound. We conclude that the quotient gets arbitrarily small as $j \to +\infty$. Thus the sequence converges, and the limit is $2 + 0 = 0$. $\square$

Insight: Another way to analyze the limit of a_j in Example 12.9 is to divide the numerator and denominator of the expression for a_j by j. This yields

$$a_j = \frac{2}{1 + 2/j}$$

Then it is apparent that $a_j \to 2/[1 + 0] = 2$.

EXAMPLE 12.10
Find the limit of the sequence

$$a_j = \frac{10^j + 8^j}{10^j}$$

Solution: If we rewrite a_j as

$$a_j = 1 + \left(\frac{8}{10}\right)^j$$

then we notice that the expression in parentheses is positive and less than 1. Higher and higher power of this number get smaller and smaller. In fact this entire expression (a number less than one raised to the jth power) tends to 0. Thus the limit of the sequence a_j exists and equals 1. □

12.2.1 SEQUENCES WITH AND WITHOUT PATTERNS

Recall that in Example 12.7 we considered the sequence $a_j = 1/j$. The pattern (or *rule*) for this sequence is a simple one. The rule for the sequence in Example 12.4 is less obvious: $f(j) = (-1)^j$. Sometimes a sequence will come from an obvious pattern or rule, and sometimes not.

EXAMPLE 12.11
What is the next element of the sequence

$$6, 6, 1, 7, 10, 2, 5, 3, 2, 5, 3$$

Solution: There is no single answer to this question. Given any finite sequence of real numbers, there are infinitely many "patterns"or "rules"which will generate them. The rule which we used was to look at the sentence immediately preceding Example 12.3 in this section of the book. The first word has six letters, the second six letters, the third one letter, and so on. The twelfth element of the sequence (the one requested above) is 9 since the twelfth word of the sentence has nine letters. The fifteenth element of the sequence is 8. Since the sentence only has fifteen words, we declare all subsequent elements of the sequence to be 0. □

Most of the sequences which we encounter in practice come from some mathematical pattern, although the pattern may be subtle.

EXAMPLE 12.12
Find the pattern in the sequence

$$2, \frac{1}{2}, \frac{4}{3}, \frac{3}{4}, \frac{6}{5}, \frac{5}{6}, \dots$$

and find the limit.

Solution: If the sequence is rewritten as

$$1 + \frac{1}{1}, 1 - \frac{1}{2}, 1 + \frac{1}{3}, 1 - \frac{1}{4}, 1 + \frac{1}{5}, \dots$$

or

$$1 + (-1)^{1+1} \cdot \frac{1}{1}, 1 + (-1)^{2+1} \cdot \frac{1}{2}, 1 + (-1)^{3+1} \cdot \frac{1}{3}, 1 + (-1)^{4+1} \cdot \frac{1}{4}, \ldots$$

then we see that the jth term is given by

$$a_j = 1 + (-1)^{j+1} \cdot \frac{1}{j}$$

Of course $1/j$ tends to 0. Therefore $\lim_{j \to \infty} a_j = 1$. ☐

12.3 The Tail of a Sequence

Intuitively, we say that $\{a_j\}_{j=1}^{\infty}$ converges to ℓ if a_j is as close as we please to ℓ when j is large enough. A crucial feature of this idea is that convergence depends only on what $\{a_j\}_{j=1}^{\infty}$ does when j is large. The values of the first 10,000 or so a_js are irrelevant.

EXAMPLE 12.13
Find the limit of the sequence defined by

$$a_j = \begin{cases} 0 & \text{if } 1 \le j \le 10,000 \\ 10 & \text{if } j \ge 10,001 \end{cases}$$

Solution: The sequence $\{a_j\}_{j=1}^{\infty}$ converges to 10. Because when j is large ($j > 10,000$), then $a_j = 10$. Far enough out, the sequence is simply constant. It has constant value and limit equal to 10. ☐

12.4 A Basic Theorem

Now we want to consider some arithmetic properties of sequences.

Theorem 12.1 *Suppose that $\{a_j\}_{j=1}^{\infty}$ and $\{b_j\}_{j=1}^{\infty}$ are convergent sequences. Then*

1. $\lim_{j \to \infty}(a_j \pm b_j) = \lim_{j \to \infty} a_j \pm \lim_{j \to \infty} b_j$
2. $\lim_{j \to \infty} (a_j \cdot b_j) = (\lim_{j \to \infty} a_j) \cdot (\lim_{j \to \infty} b_j)$

3. $\lim_{j\to\infty} \frac{a_j}{b_j} = \frac{\lim_{j\to\infty} a_j}{\lim_{j\to\infty} b_j}$ *provided that* $\lim_{j\to\infty} b_j \neq 0$

4. $\lim_{j\to\infty} \alpha \cdot a_j = \alpha \cdot \lim_{j\to\infty} a_j$ *for any real number α*

5. $\lim_{j\to\infty} a_j$ *is unique*

Insight: In what follows we will generally not mention Part 5 explicitly, but we use it frequently. It means that if we calculate a limit by any particular method, the answer will be the same as if some other method were used instead: there is no ambiguity in the theory of limits.

EXAMPLE 12.14
Compute

$$\lim_{j\to\infty} \left(\frac{3}{j} + 1 \right)$$

Solution: We apply the theorem several times as follows:

(Part 1) $\lim_{j\to\infty} \left(\frac{3}{j} + 1 \right) = \lim_{j\to\infty} \frac{3}{j} + \lim_{j\to\infty} 1$

(Part 4) $= 3 \cdot \lim_{j\to\infty} \frac{1}{j} + \lim_{j\to\infty} 1$

(Example 12.7) $= 3 \cdot 0 + \lim_{j\to\infty} 1$

(Example 12.8) $= 1$ $\square$

Insight: In the beginning, you should work through problems step-by-step, carefully noting each time you apply one of the parts of Theorem 12.1. After a while, this meticulousness will be unnecessary and you will be able to do many problems in your head.

EXAMPLE 12.15
Compute

$$\lim_{j\to\infty} \frac{1/j + 1}{3/j^2 - 5}$$

Solution: We systematically apply Theorem 1 to find that the limit equals

(Part 3) $\dfrac{\lim_{j\to\infty}(1/j + 1)}{\lim_{j\to\infty}(3/j^2 - 5)}$

(Part 1) $= \dfrac{\lim_{j\to\infty} 1/j + \lim_{j\to\infty} 1}{\lim_{j\to\infty} 3/j^2 - \lim_{j\to\infty} 5}$

(Parts 2, 4)
$$= \frac{\lim_{j\to\infty} 1/j + \lim_{j\to\infty} 1}{3 \cdot \lim_{j\to\infty} 1/j \cdot \lim_{j\to\infty} 1/j - \lim_{j\to\infty} 5}$$

(Examples 12.7, 12.8)
$$= \frac{0+1}{3 \cdot 0 \cdot 0 - 5}$$

$$= -\frac{1}{5} \qquad \qquad \square$$

EXAMPLE 12.16

Use Theorem 12.1 to analyze the limit

$$\lim_{j\to\infty} \frac{j^3 - 4j + 6}{3j^3 + 2j}$$

Solution: Notice that the limit cannot be handled directly using Theorem 12.1 since both the numerator and the denominator become ever larger with j (they tend to no limit). It is not entirely clear that the limit exists! The approach we take is to use some algebra before we attempt to apply Theorem 12.1.

Dividing numerator and denominator by the largest power of j that appears (namely, j^3), we find that our limit equals

$$\lim_{j\to\infty} \frac{1 - 4j^{-2} + 6j^{-3}}{3 + 2j^{-2}}$$

(Part 3)
$$= \frac{\lim_{j\to\infty}(1 - 4j^{-2} + 6j^{-3})}{\lim_{j\to\infty}(3 + 2j^{-2})}$$

(Parts 1,4)
$$= \frac{\lim_{j\to\infty} 1 - 4 \cdot \lim_{j\to\infty} j^{-2} + 6 \cdot \lim_{j\to\infty} j^{-3}}{\lim_{j\to\infty} 3 + 2 \lim_{j\to\infty} j^{-2}}$$

However, using Part 1 of Theorem 12.1 and Example 12.7, we see that

$$\lim_{j\to\infty} j^{-2} = \lim_{j\to\infty} j^{-1} \cdot \lim_{j\to\infty} j^{-1} = 0 \cdot 0 = 0$$

A similar calculation shows that

$$\lim_{j\to\infty} j^{-3} = 0$$

We conclude that our limit equals

$$\frac{1 - 4 \cdot 0 + 6 \cdot 0}{3 + 2 \cdot 0} = \frac{1}{3} \qquad \qquad \square$$

From now on, in order to facilitate the flow of ideas, we will not continue to cite each use of Theorem 1. But as you read the examples you should check for yourself why each step is correct.

You Try It: Analyze the sequence $a_j = \sqrt{j}/[\sqrt{j} + 5]$.

12.5 The Pinching Theorem

We turn now to a "pinching theorem" for sequences. This tool gives us a means for comparing a new, perhaps difficult and mysterious, sequence with other more familiar sequences.

Theorem 12.2 (The Pinching Theorem) *Suppose that* $\{a_j\}_{j=1}^{\infty}, \{b_j\}_{j=1}^{\infty}, \{c_j\}_{j=1}^{\infty}$ *are sequences. If*

$$a_j \leq b_j \leq c_j \qquad \text{for every } j$$

and if

$$\lim_{j \to \infty} a_j = \lim_{j \to \infty} c_j = \ell$$

then

$$\lim_{j \to \infty} b_j = \ell$$

EXAMPLE 12.17
Evaluate

$$\lim_{j \to \infty} \frac{(\sin j)^2}{j}$$

Solution: Let $b_j = \frac{(\sin j)^2}{j}$ Then

$$0 \le b_j \le \frac{1}{j}$$

To apply the pinching theorem, let

$$a_j = 0 \quad \text{and} \quad c_j = \frac{1}{j} \quad \text{for every} \quad j$$

We observe that

$$\lim_{j \to \infty} a_j = \lim_{j \to \infty} c_j = 0$$

The hypotheses of the pinching theorem are satisfied and we may conclude that

$$\lim_{j \to \infty} \frac{(\sin j)^2}{j} = \lim_{j \to \infty} b_j = 0 \qquad \square$$

12.6 Some Special Sequences

A number of special sequences occur repeatedly in our work. We now collect several of them for easy reference.

Theorem 12.3 *Let s be a real number.*

1. *If $s < 0$ then the sequence $\{j^s\}_{j=1}^{\infty}$ converges to 0.*
2. *If $s > 0$ then the sequence $\{j^s\}_{j=1}^{\infty}$ diverges.*
3. *If $s = 0$ then the sequence $\{j^s\}_{j=1}^{\infty}$ is just the constant sequence $1, 1, 1, \ldots$ and converges to 1.*

Theorem 12.4 is a companion to Theorem 12.3. In Theorem 12.3 the j's are the base of the exponential, while in Theorem 12.4 we have j playing the role of the exponent.

Theorem 12.4 *Let t be a real number.*

1. *If $|t| < 1$ then $\{t^j\}_{j=1}^{\infty}$ converges to 0.*
2. *If $|t| > 1$ then $\{t^j\}_{j=1}^{\infty}$ diverges.*

3. *If $t = 1$ then $\{t^j\}_{j=1}^{\infty}$ is the constant sequence $1, 1, 1, \ldots$ which converges to 1.*

4. *If $t = -1$ then $\{t^j\}_{j=1}^{\infty}$ diverges.*

EXAMPLE 12.18

Calculate

$$\lim_{j \to \infty} \frac{j^{3/2} + 2j^2}{4j^2}$$

Solution: We use some algebraic manipulations to rewrite our limit as

$$\lim_{j \to \infty} \frac{j^{3/2}}{4j^2} + \lim_{j \to \infty} \frac{2j^2}{4j^2}$$

$$= \frac{1}{4} \cdot \lim_{j \to \infty} \frac{1}{j^{1/2}} + \lim_{j \to \infty} \frac{1}{2}$$

By Theorem 12.3 this last line equals

$$\frac{1}{4} \cdot 0 + \frac{1}{2} = \frac{1}{2} \qquad \qquad \square$$

EXAMPLE 12.19

Calculate

$$\lim_{j \to \infty} \frac{2^j + 3^j}{4^j}$$

Solution: We multiply the numerator and denominator by 4^{-j} and simplify to obtain that our limit equals

$$\lim_{j \to \infty} \frac{2^j \cdot 4^{-j} + 3^j \cdot 4^{-j}}{4^j \cdot 4^{-j}} = \lim_{j \to \infty} \frac{(1/2)^j + (3/4)^j}{1}$$

$$= \lim_{j \to \infty} \left(\frac{1}{2}\right)^j + \lim_{j \to \infty} \left(\frac{3}{4}\right)^j$$

Now by Theorem 12.4(1) this equals

$$0 + 0 = 0 \qquad \qquad \square$$

Exercises

In Exercises 1 and 2, use your intuition to determine whether the given sequence converges. If it does, say what the limit is. *Do not attempt to supply proofs.* Use your calculator if you wish to gather data.

1. $a_j = \frac{1}{j^2+1}$

2. $a_j = \frac{j}{j^2+1}$

In Exercises 3 and 4, determine intuitively whether the limit exists and what the limit is. *Then* verify your answer rigorously using the precise definition of limit.

3. $\lim_{j\to\infty} \frac{1}{j+7}$

4. $\lim_{j\to\infty} 10^{-j}$

It is important for you to realize that there are many sequences whose convergence properties are not at all obvious. Each of the sequences in Exercises 5 and 6 converges, but it is quite tricky to determine what the limit is and to prove the answer. Use a calculator to help you guess what the limit is in each problem, but do not worry for now about proving your answer rigorously.

5. $\lim_{j\to\infty} j \cdot \sin(1/j)$

6. $\lim_{j\to\infty} j^2 \cdot (1 - \cos(1/j))$

In Exercises 7 and 8, compute the limit using the rules which you learned in Theorems 12.1, 12.3, and 12.4. Note explicitly each time that you use a rule.

7. $\lim_{j\to\infty} \left(1 + \frac{1}{j}\right)$

8. $\lim_{j\to\infty} \left(\frac{1}{j} - \frac{1}{j^2}\right)$

Use the pinching theorem to evaluate the limits in Exercises 9 and 10.

9. $\lim_{j\to\infty} \frac{\sin(1/j)}{j}$

10. $\lim_{j\to\infty} \frac{\sqrt{j^4+1}}{j^2}$

In Exercises 11, compute the limit using the rules which you learned in this chapter. Note explicitly each time that you use a rule.

11. $\lim_{j\to\infty}(2^{-j} + 3)$

CHAPTER 13

Series

13.1 Fundamental Ideas

The idea of adding together finitely many numbers is a simple and familiar one. We are so used to the associative and commutative properties of finite sums that we tend not to think about them. Now we shall learn to add together *infinitely many* numbers, and we will have to be a bit more careful. As Example 13.1 suggests, contradictions can arise if we do not establish certain ground rules.

EXAMPLE 13.1
Give an example of an infinite summation process in which the ordinary rules of arithmetic appear to fail.

Solution: Consider the infinite sum

$$1 - 1 + 1 - 1 + 1 - 1 + 1 - 1 + \cdots$$

If we group the terms as

$$(1 - 1) + (1 - 1) + (1 - 1) + (1 - 1) + \cdots$$

then the sum seems to be

$$0 + 0 + 0 + 0 + \cdots$$

which ought to be 0. On the other hand, if we group the terms as

$$1 + (-1 + 1) + (-1 + 1) + (-1 + 1) + \cdots$$

then the sum seems to be

$$1 + 0 + 0 + 0 + \cdots$$

which ought to be 1.

Apparently the associative law of addition fails for this infinite sum. □

It turns out that, while the associative law is valid for finite sums, it is not valid *in general* for infinite sums. In order to make sense of this and other addition operations, we must first consider precisely what it means to "add" infinitely many numbers.

We begin by recalling the following notation from Sec. 6.4:

Definition 13.1 The symbol

$$\sum_{j=M}^{N} c_j$$

denotes the sum of the numbers $c_M, c_{M+1}, \ldots, c_{N-1}, c_N$. In other words

$$\sum_{j=M}^{N} c_j = c_M + c_{M+1} + \cdots + c_{N-1} + c_N$$

EXAMPLE 13.2
If $c_j = 2^{-j}$ then calculate $\sum_{j=1}^{4} c_j$ and $\sum_{j=3}^{7} c_j$.

Solution: We have

$$\sum_{j=1}^{4} c_j = 2^{-1} + 2^{-2} + 2^{-3} + 2^{-4} = \frac{15}{16}$$

and

$$\sum_{j=3}^{7} c_j = 2^{-3} + 2^{-4} + 2^{-5} + 2^{-6} + 2^{-7} = \frac{31}{128} \qquad \square$$

Definition 13.2 If $c_1, c_2, c_3, \ldots$ are real numbers and $\sum_{j=1}^{\infty} c_j$ the corresponding series, then we write S_N to denote the sum of the first N of them:

$$S_N = \sum_{j=1}^{N} c_j$$

We call S_N the Nth *partial sum* of the *series* $\sum_{j=1}^{\infty} c_j$.
   If

$$\lim_{N \to \infty} S_N = \ell$$

then we say that the infinite sum (or *series*) $\sum_{j=1}^{\infty} c_j$ *converges* to ℓ. If the sequence $\{S_N\}_{N=1}^{\infty}$ does not converge then we say that the series *diverges*.

Insight: The idea in this definition is as follows: to add up infinitely many numbers $c_1 + c_2 + c_3 + \cdots$ we add up a large finite number of the terms. That is, we consider the partial sums

$$S_N = c_1 + c_2 + c_3 + \cdots + c_N$$

In the case that the series converges, this partial sum should give us a good approximation to the full sum $\sum_{j=1}^{\infty} c_j$.
   In practice, when we write

$$\sum_{j=1}^{\infty} c_j$$

we mean

$$\lim_{N \to \infty} S_N$$

provided that this limit exists. $\square$

13.2 Some Examples

EXAMPLE 13.3
Does the series $\sum_{j=1}^{\infty} 2^{-j}$ converge? If so, to what limit?

Solution: We have

$$S_N = \sum_{j=1}^{N} 2^{-j} = 2^{-1} + 2^{-2} + 2^{-3} + \cdots + 2^{-N}$$

$$= 2^{-1} + (2^{-1} - 2^{-2}) + (2^{-2} - 2^{-3}) + (2^{-3} - 2^{-4})$$

$$+ \cdots + (2^{-(N-1)} - 2^{-N}) \tag{13.1}$$

Notice that S_N is a finite sum so that it is correct to use the associative law of addition. We conclude that

$$S_N = 2^{-1} + 2^{-1} + (-2^{-2} + 2^{-2}) + (-2^{-3} + 2^{-3})$$

$$+ \cdots + (-2^{-(N-1)} + 2^{-(N-1)}) - 2^{-N}$$

$$= 1 - 2^{-N}$$

We may now pass to the limit:

$$\lim_{N \to \infty} S_N = \lim_{N \to \infty} (1 - 2^{-N}) = 1 - \lim_{N \to \infty} 2^{-N} = 1 - 0 = 1$$

By definition, we say that the series

$$\sum_{j=1}^{\infty} 2^{-j}$$

converges to 1. Refer to Fig. 13.1. □

Figure 13.1 Convergence of the series $\sum_{j} 2^{-j}$.

Insight: Do not worry about how one thinks of the trick that was used in Example 13.3. The purpose of this chapter is to teach you all the techniques that you will need in order to study series. We will learn them gradually, beginning in the next section. For now, just concentrate on learning what it means for a series to converge or diverge.

EXAMPLE 13.4
Discuss convergence for the series

$$\sum_{j=1}^{\infty}(-1)^{j+1} = 1 - 1 + 1 - 1 + 1 - \cdots$$

Solution: If N is odd then

$$S_N = 1 - 1 + 1 - 1 + - \cdots + 1$$
$$= (1 - 1) + (1 - 1) + \cdots + (1 - 1) + 1$$
$$= 1$$

However, if N is even then

$$S_N = 1 - 1 + 1 - 1 + - \cdots - 1$$
$$= (1 - 1) + (1 - 1) + \cdots + (1 - 1)$$
$$= 0$$

Therefore the sequence $\{S_N\}_{N=1}^{\infty}$ of partial sums is just the sequence

$$1, 0, 1, 0, 1, 0, \ldots$$

which does not converge.
   We conclude that the series itself does not converge.   □

Insight: Example 13.4 explains away the apparent contradiction that we encountered in Example 13.1. The lesson is that we should not attempt to perform arithmetic operations on series that do not converge.

WARNING: It is easy to become confused at this point about the difference between a sequence and a series. Remember that a *sequence* is a *list* of numbers while a *series* is a *sum* of numbers. However, we study a series by looking at its sequence of partial sums.

Sometimes there are tricks involved in seeing that a series converges:

EXAMPLE 13.5
Discuss convergence for the series

$$\sum_{j=1}^{\infty} \frac{1}{j \cdot (j+1)}$$

Solution: If you use a calculator to compute some partial sums (up to S_{20}, for instance) then you will probably be convinced that the series converges. Here is a way to get this conclusion mathematically: We write

$$S_N = \frac{1}{1 \cdot 2} + \frac{1}{2 \cdot 3} + \frac{1}{3 \cdot 4} + \cdots + \frac{1}{N \cdot (N+1)}$$

$$= \left(\frac{1}{1} - \frac{1}{2}\right) + \left(\frac{1}{2} - \frac{1}{3}\right) + \left(\frac{1}{3} - \frac{1}{4}\right) + \cdots + \left(\frac{1}{N} - \frac{1}{N+1}\right)$$

Almost everything cancels out and we have

$$S_N = 1 - \frac{1}{N+1}$$

Clearly $\lim_{N \to \infty} S_N = 1$. Therefore the series

$$\sum_{j=1}^{\infty} \frac{1}{j \cdot (j+1)}$$

converges to 1. □

Insight: Sometimes it is convenient to begin a series at an index other than $j = 1$. An example is

$$\sum_{j=3}^{\infty} \frac{j}{\ln(j-1)}$$

Obviously it would not do to begin this series at $j = 1$ because $\ln 0$ is undefined. We cannot begin at $j = 2$ because $\ln 1 = 0$. So we begin at $j = 3$. Of course this

series is equal to

$$\frac{3}{\ln 2} + \frac{4}{\ln 3} + \cdots$$

But notice this: the partial sum S_1 is understood to be 0 because summing from 3 to 1 makes no sense. The partial sum S_2 is zero for a similar reason. The first nontrivial partial sum is

$$S_3 = \frac{3}{\ln 2}$$

Indeed, for $N \geq 3$ we have

$$S_N = \frac{3}{\ln 2} + \frac{4}{\ln 3} + \cdots + \frac{N}{\ln(N-1)}$$

EXAMPLE 13.6
What are S_5, S_{10}, and S_4 for the series

$$\sum_{j=5}^{\infty} 3^j$$

Solution: The interest of this example is that the series does not begin at $j = 1$. We have

$$S_5 = 3^5$$

$$S_{10} = 3^5 + 3^6 + 3^7 + 3^8 + 3^9 + 3^{10}$$

$$S_4 = 0 \qquad\qquad \square$$

13.3 The Harmonic Series

Example 13.5 was devious, but it introduced a useful technique. In the next example we consider a series that *diverges* in a subtle way. This is the important *harmonic series*.

EXAMPLE 13.7
Discuss convergence or divergence for the series

$$\sum_{j=1}^{\infty} \frac{1}{j}$$

Solution: If you calculate a large number of (for instance several hundred) partial sums using a computer then you might never suspect that this series diverges. It turns out that the sum of the first one million terms of this series is just under 14. The divergence is taking place so slowly that it is difficult to be certain what is actually happening. We need a mathematical proof:

Notice that

$$S_1 = 1 = \frac{2}{2}$$

$$S_2 = 1 + \frac{1}{2} = \frac{3}{2}$$

$$S_4 = 1 + \frac{1}{2} + \left(\frac{1}{3} + \frac{1}{4}\right)$$

$$\geq 1 + \frac{1}{2} + \left(\frac{1}{4} + \frac{1}{4}\right) \geq 1 + \frac{1}{2} + \frac{1}{2} = \frac{4}{2}$$

$$S_8 = 1 + \frac{1}{2} + \left(\frac{1}{3} + \frac{1}{4}\right) + \left(\frac{1}{5} + \frac{1}{6} + \frac{1}{7} + \frac{1}{8}\right)$$

$$\geq 1 + \frac{1}{2} + \left(\frac{1}{4} + \frac{1}{4}\right) + \left(\frac{1}{8} + \frac{1}{8} + \frac{1}{8} + \frac{1}{8}\right)$$

$$= \frac{5}{2}$$

In general this argument shows that

$$S_{2^k} \geq \frac{k+2}{2}$$

The sequence of S_N's is increasing since the series contains only positive terms. The fact that the partial sums $S_1, S_2, S_4, S_8, \ldots$ increase without bound shows that the entire sequence of partial sums must increase without bound. We conclude that the series diverges. □

Insight: The series $\sum_{j=1}^{\infty} \frac{1}{j}$ comes up often in mathematics and physics. It is called the "harmonic series" because of its role in the theory of acoustics: if the natural frequencies at which a given body vibrates are $\rho_1, \rho_2, \ldots$ and if

$$\frac{\rho_j}{\rho_{j+1}} = \frac{1}{j+1}$$

for all j then the frequencies are said to form a sequence of harmonics.

13.4 Series of Powers

We now consider series of powers of a fixed number. These are called *geometric series*.

EXAMPLE 13.8

Discuss convergence for the series $\sum_{j=1}^{\infty} 10^{-j}$.

Solution: Notice that

$$S_1 = 0.1$$
$$S_2 = 0.11$$
$$S_3 = 0.111$$

and so on

We see that the partial sums form an increasing sequence that is bounded above by 1. It is plausible—and there is a mathematical theorem that makes this assertion rigorous—that such a sequence must accumulate at some point less than or equal to 1. Thus the series converges. (We shall treat repeating decimals in more detail in the next section.) To what number does the series converge?

Referring back to the material in Sec. 5.4 on the real number system, we see that the sequence of partial sums $\{S_N\}_{N=1}^{\infty}$ converges to the real number given by the infinite decimal expansion

$$0.11111\ldots = \frac{1}{9}$$

Therefore the series $\sum_{j=1}^{\infty} 10^{-j}$ converges to 1/9 .   □

EXAMPLE 13.9
Discuss convergence for the series $\sum_{j=1}^{\infty} 17^{-j}$.

Solution: Notice that

$$S_1 < S_2 < S_3 < \cdots$$

since all of the summands are positive. Also

$$
\begin{aligned}
S_N &= 17^{-1} + 17^{-2} + 17^{-3} + \cdots + 17^{-N} \\
&< 2^{-1} + 2^{-2} + 2^{-3} + \cdots + 2^{-N} \\
&= 1 - 2^{-N} \\
&< 2
\end{aligned}
$$

(We know this from the calculation that we did in Example 13.3). Therefore the sequence of partial sums $S_1, S_2, S_3, \ldots$ is increasing and bounded above by 2. As noted before, when a sequence increases and is bounded above, it must accumulate at some point. Therefore our partial sums have a limit ℓ. We conclude that the series converges. □

Insight: Notice in the last example that we demonstrated that the series converges *without actually finding the sum*. This is the way matters will often turn out in our study of series.

13.5 Repeating Decimals

In this section we begin to understand infinitely repeating decimals. The way we do this is that we learn to convert them to rational fractions.

EXAMPLE 13.10
Consider the real number given by the repeating decimal expansion

$$x = 2.713131313\ldots$$

We often find it convenient to write a number like this as

$$x = 2.7\overline{13}$$

where the overbar indicates that the decimal digits under it are repeated indefinitely. This number x is a legitimate real number, but we would like to write it in a more compact (or perhaps more understandable) form.

The device that we use is to compare $100x$ with x. Thus we have

$$100x = 271.3131313\ldots$$

$$x = 2.7131313\ldots$$

Clearly the multiple of 100 was chosen so that the infinite repetions of 13 line up nicely in our array. Subtracting, we find that

$$99x = 268.6$$

or

$$x = \frac{268.6}{99} = \frac{2686}{990}$$

As a result of our calculation, we have expressed the infinitely repeating decimal x as a rational fraction. $\square$

You Try It: Use the technique just presented to convert the repeating decimal

$$0.11111\ldots$$

to a rational fraction. Of course you should obtain the answer 1/9.

EXAMPLE 13.11
Let us convert the repeating decimal

$$y = 37.12712712\ldots = 37.12\overline{712}$$

to a rational fraction.
   We write

$$1000y = 37127.127127127\ldots$$

$$y = 37.127127127\ldots$$

Subtracting, we find that

$$999y = 37090$$

or

$$y = \frac{37090}{999}$$

□

It is an interesting fact that decimal expansions that repeat—such as the ones we considered in the last two examples—*always* give rise to rational numbers (that is, quotients of integers). Decimal expansions that *do not* repeat always give rise to irrational numbers (that is, numbers that *cannot* be expressed as quotients of integers).

13.6 An Application

In this section we illustrate some of our ideas about series with an example from the physical world.

EXAMPLE 13.12
A rocket uses a great deal of fuel during liftoff. Once the rocket leaves the earth's gravitational field, relatively little fuel is required to keep the rocket in motion. Scientists estimate that, for a certain class of rocket, the vehicle uses 1/4 of its total fuel supply during its first hour of flight, 1/8 of its fuel supply during the second hour, 1/12 of its fuel supply during the third hour, and so on. A newspaper reporter hears this information and reports that the rocket's fuel supply will "last forever." Why is the reporter in error?

Solution: The portion of the total fuel supply that the rocket has consumed after N hours is

$$\frac{1}{4} + \frac{1}{8} + \frac{1}{12} + \cdots + \frac{1}{4N}$$

This is a partial sum S_N for the series

$$\sum_{j=1}^{\infty} \frac{1}{4j}$$

Thus

$$S_N = \frac{1}{4}\left(\frac{1}{1} + \frac{1}{2} + \frac{1}{3} + \cdots + \frac{1}{N}\right)$$

But we learned in Example 13.7 that the expression in parentheses becomes large without bound as N increases. In particular, $S_N > 1$ when N is large enough, so the rocket will eventually exhaust its fuel supply (in fact a calculation shows that the fuel will be exhausted during the thirty-first hour of flight). □

13.7 A Basic Test for Convergence

Theorem 13.1 *(The zero test) If the series $\sum_{j=1}^{\infty} c_j$ converges then the summands c_j tend to* 0.

Proof: Since $\sum_{j=1}^{\infty} c_j$ converges, the partial sums S_N tend to some limit ℓ. Let $\epsilon > 0$. Choose N_0 such that if $N \geq N_0$ then $|S_N - \ell| < \epsilon/2$. Then also $|S_{N+1} - \ell| < \epsilon/2$. Therefore

$$|c_{N+1} - 0| = |c_{N+1}| = |S_{N+1} - S_N|$$
$$= |(S_{N+1} - \ell) + (\ell - S_N)|$$
$$\leq |S_{N+1} - \ell| + |\ell - S_N|$$
$$< \frac{\epsilon}{2} + \frac{\epsilon}{2}$$
$$= \epsilon$$

This says that $c_N \to 0$. □

EXAMPLE 13.13
Apply the zero test to the series $\sum_{j=1}^{\infty} (-1)^j$.

Solution: The summands $c_j = (-1)^j$ do not tend to zero therefore the series diverges. □

EXAMPLE 13.14
Apply the zero test to the series $\sum_{j=1}^{\infty} \frac{1}{j}$.

Solution: The terms $c_j = 1/j$ certainly tend to 0. But we cannot conclude that the series converges (that is *not* what Theorem 13.1 says). In fact we proved in the last section that the series diverges. □

EXAMPLE 13.15
Apply the zero test to the series $\sum_{j=1}^{\infty} 17^{-j}$.

Solution: The terms $c_j = 17^{-j}$ certainly tend to zero. So the series *might* converge, but the zero test will not tell us. In Example 13.9, we in fact learned that the series *does* converge. □

13.8 Basic Properties of Series

We now present some elementary properties of series that are similar to the properties of sequences given in Theorem 12.1.

Theorem 13.2 *Suppose that $\sum_{j=1}^{\infty} c_j$ converges to C and $\sum_{j=1}^{\infty} d_j$ converges to D. Then*

1. $\sum_{j=1}^{\infty}(c_j \pm d_j)$ *converges to $C \pm D$.*
2. $\sum_{j=1}^{\infty} \alpha c_j$ *converges to $\alpha \cdot C$, where α is any real constant.*

Corollary 13.1 *If $\sum_{j=1}^{\infty} b_j$ diverges and α is any nonzero constant then $\sum_{j=1}^{\infty} \alpha \cdot b_j$ diverges.*

Proof of Corollary 13.1: Seeking a contradiction, we suppose that $\sum_{j=1}^{\infty} \alpha \cdot b_j$ converges. By Part (2) of the theorem,

$$\sum_{j=1}^{\infty} \frac{1}{\alpha} \cdot \alpha \cdot b_j = \sum_{j=1}^{\infty} b_j$$

also converges. That is a contradiction. □

Here are some examples which indicate how Theorem 13.2 and its corollary can be applied.

EXAMPLE 13.16
Discuss convergence for the series

$$\sum_{j=1}^{\infty} \frac{4}{2^j} \tag{13.2}$$

Solution: We know by Example 13.3 that

$$\sum_{j=1}^{\infty} \frac{1}{2^j}$$

converges. But then Part (2) of Theorem 13.2 applies to tell us that the series in Eq. (13.2) converges.   □

EXAMPLE 13.17
Does the series

$$\sum_{j=1}^{\infty} \frac{2^j + 10^j}{20^j}$$

converge?

Solution: We rewrite the jth summand as

$$\frac{2^j + 10^j}{20^j} = \frac{1}{10^j} + \frac{1}{2^j}$$

Because

$$\sum_{j=1}^{\infty} \frac{1}{2^j}$$

converges (Example 13.3) and

$$\sum_{j=1}^{\infty} \frac{1}{10^j}$$

converges (Example 13.8), we may apply Part (1) of Theorem 13.2 to conclude that

$$\sum_{j=1}^{\infty} \left(\frac{1}{10^j} + \frac{1}{2^j} \right)$$

converges. Therefore the original series converges.   □

EXAMPLE 13.18
Does the series

$$\sum_{j=1}^{\infty} \frac{j + 2^j}{j \cdot 2^j} \tag{13.3}$$

converge?

Solution: Suppose that the series does converge. We also know that

$$\sum_{j=1}^{\infty} \frac{1}{2^j}$$

converges (by Example 13.3). Therefore Part (1) of Theorem 13.2 tells us that

$$\sum_{j=1}^{\infty} \left(\frac{j + 2^j}{j \cdot 2^j} - \frac{1}{2^j} \right)$$

converges.

Performing the subtraction, we may conclude that

$$\sum_{j=1}^{\infty} \frac{1}{j}$$

converges. But that is false (Example 13.7). Therefore our assumption is contradicted and the original series [Eq. (13.3)] diverges.   □

EXAMPLE 13.19
Discuss convergence of the series

$$\sum_{j=1}^{\infty} (2^{-j+7} - 10^{-j+3})$$

Solution: Rewrite the series as

$$\sum_{j=1}^{\infty} (128 \cdot 2^{-j} - 1000 \cdot 10^{-j})$$

Now

$$\sum_{j=1}^{\infty} 2^{-j}$$

is convergent therefore Part (2) of Theorem 13.2 tells us that

$$\sum_{j=1}^{\infty} 128 \cdot 2^{-j} \tag{13.4}$$

converges. Similarly,

$$\sum_{j=1}^{\infty} 10^{-j}$$

converges hence we may conclude that

$$\sum_{j=1}^{\infty} 1000 \cdot 10^{-j} \tag{13.5}$$

converges.

Finally, Eqs. (13.4) and (13.5), together with Part (1) of Theorem 13.2, tell us that the original series converges. □

13.9 Geometric Series

We will next discuss how to sum explicitly a special kind of series called the *geometric series*. (We already began exploring these in the last section.)

Let λ be a fixed real number and consider the series $\sum_{j=0}^{\infty} \lambda^j$. We find it convenient here to begin the sum with the index $j = 0$ rather than $j = 1$. By convention, $\lambda^0 = 1$. Then

$$S_N = 1 + \lambda + \lambda^2 + \cdots + \lambda^N$$

Therefore

$$\begin{aligned}
\lambda \cdot S_N &= \lambda + \lambda^2 + \lambda^3 + \cdots + \lambda^{N+1} \\
&= (1 + \lambda + \lambda^2 + \cdots + \lambda^N) + (\lambda^{N+1} - 1) \\
&= S_N + \lambda^{N+1} - 1
\end{aligned}$$

We reorganize the equation so that the terms involving S_N are together:

$$S_N(\lambda - 1) = \lambda^{N+1} - 1$$

As long as $\lambda \neq 1$, we may divide by $(\lambda - 1)$ to obtain

$$S_N = \frac{1 - \lambda^{N+1}}{1 - \lambda} \tag{13.6}$$

This formula for S_N tells us everything that we could want to know about the geometric series $\sum_{j=0}^{\infty} \lambda^j$.

Theorem 13.3 *If* $|\lambda| < 1$ *then the geometric series* $\sum_{j=0}^{\infty} \lambda^j$ *converges to the sum* $1/(1 - \lambda)$. *If* $|\lambda| \geq 1$ *then the geometric series* $\sum_{j=0}^{\infty} \lambda^j$ *does not converge.*

Proof: Remember what it means for a series to converge: the sequence of partial sums S_N converges to some limit. If $|\lambda| < 1$ then by (13.6) we see that

$$S_N \to \frac{1 - 0}{1 - \lambda} = \frac{1}{1 - \lambda}$$

That is the first statement of the theorem.

If $|\lambda| \geq 1$ then the terms of the series do not tend to zero so that, by the zero test, the series cannot converge. $\square$

EXAMPLE 13.20
Discuss convergence for the series

$$\sum_{j=0}^{\infty} 3^{-j}$$

Solution: We rewrite the series as

$$\sum_{j=0}^{\infty} \left(\frac{1}{3}\right)^j$$

which converges to

$$\frac{1}{1 - (1/3)} = \frac{3}{2}$$

(because $|1/3| < 1$). $\square$

EXAMPLE 13.21
Discuss convergence for the series

$$\sum_{j=0}^{\infty} (-2)^j$$

Solution: Since $|-2| > 1$, the series must diverge. $\square$

EXAMPLE 13.22
Sum the series

$$\sum_{k=4}^{\infty} 6^{-k} \tag{13.7}$$

explicitly.

Solution: We begin by rewriting the series as

$$\sum_{k=0}^{\infty} 6^{-k-4} \tag{13.8}$$

Why can we do this? Write out the first several terms of both Eqs. (13.7) and (13.8) to see that they are simply two different way of writing

$$6^{-4} + 6^{-5} + 6^{-6} + \cdots$$

Now we may use Theorem 13.2(1) of the last section to rewrite (13.8) as

$$6^{-4} \cdot \sum_{k=0}^{\infty} 6^{-k} = 6^{-4} \cdot \sum_{k=0}^{\infty} \left(\frac{1}{6}\right)^k$$

Finally,

$$\sum_{k=0}^{\infty} \left(\frac{1}{6}\right)^k = \frac{1}{1 - (1/6)} = \frac{6}{5}$$

We conclude that

$$\sum_{k=4}^{\infty} 6^{-k} = 6^{-4} \cdot \frac{6}{5} = \frac{1}{1080} \qquad \square$$

EXAMPLE 13.23
Sum the series

$$\sum_{j=6}^{\infty} \frac{2^{j+4}}{7^{j-3}}$$

explicitly.

Solution: We rewrite the series as

$$\sum_{j=0}^{\infty} \frac{2^{j+10}}{7^{j+3}} = \frac{2^{10}}{7^3} \cdot \sum_{j=0}^{\infty} \frac{2^j}{7^j}$$

$$= \frac{1024}{343} \cdot \sum_{j=0}^{\infty} \left(\frac{2}{7}\right)^j$$

By Theorem 13.3,

$$\sum_{j=0}^{\infty} \left(\frac{2}{7}\right)^j = \frac{1}{1 - (2/7)}$$

$$= \frac{7}{5}$$

Therefore

$$\sum_{j=6}^{\infty} \frac{2^{j+4}}{7^{j-3}} = \frac{1024}{343} \cdot \frac{7}{5} = \frac{1024}{245} \qquad \Box$$

Insight: Whether or not a given series converges does not depend on the first million or so terms. For suppose that we are given the series $\sum_{j=1}^{\infty} a_j$ and that we are able to ascertain that

$$\sum_{j=10^6}^{\infty} a_j$$

converges. If $N \geq 10^6$ then this means that the partial sums

$$S_N = \sum_{j=10^6}^{N} a_j$$

converge (as a sequence) to a limit ℓ. But then the expressions

$$T_N = \sum_{j=1}^{N} a_j = \sum_{j=1}^{10^6-1} a_j + \sum_{j=10^6}^{N} a_j = \sum_{j=1}^{10^6-1} a_j + S_N$$

converge as $N \to \infty$ to

$$\sum_{j=1}^{10^6-1} a_j + \ell$$

(Notice that the first sum does not depend on N.) What is T_N? It is the partial sum for the full series

$$\sum_{j=1}^{\infty} a_j$$

In other words, we have concluded that the full series converges. To summarize:

If $\sum_{j=10^6}^{\infty} a_j$ converges then $\sum_{j=1}^{\infty} a_j$ converges.

Conversely,

If $\sum_{j=1}^{\infty} a_j$ converges then $\sum_{j=10^6}^{\infty} a_j$ converges.

The reason is that the two sums differ only by the finite number

$$\sum_{j=1}^{10^6-1} a_j \qquad \qquad \Box$$

EXAMPLE 13.24
Does the series

$$\sum_{j=75}^{\infty} (1.1)^j$$

converge?

Solution: The series must diverge. For if it converged, then it would follow that $\sum_{j=0}^{\infty}(1.1)^j$ converges. That, of course, is false since $|1.1| > 1$ $\qquad \Box$

You Try It: Discuss convergence for the series $\sum_j (0.9)^{2j}$.

EXAMPLE 13.25
Does the series

$$\sum_{j=20}^{\infty} (0.9)^j$$

converge?

Solution: Yes. For

$$\sum_{j=0}^{\infty} (0.9)^j$$

converges (because $|0.9| < 1$) and the two series differ by just the finite sum

$$0.9^0 + 0.9^1 + 0.9^2 + \cdots + 0.9^{19} \qquad \square$$

EXAMPLE 13.26
Calculate the sum

$$\sum_{j=7}^{20} 5 \cdot 4^j$$

explicitly.

Solution: Notice that the limits of summation are not what we are used to; in particular, the lower limit is not 0. We rewrite our sum as

$$5 \cdot \left(\sum_{j=0}^{20} 4^j - \sum_{j=0}^{6} 4^j \right)$$

Now we may use the formula for partial sums of geometric series which appears in line (13.6). We find that the sum equals

$$5 \cdot \left(\frac{1 - 4^{21}}{1 - 4} - \frac{1 - 4^7}{1 - 4} \right)$$

Of course this solution can be simplified, if necessary, with a little further calculation. $\qquad \square$

13.9.1 AN APPLICATION

EXAMPLE 13.27
At a certain aluminum-recycling plant, the recycling process turns n pounds of used aluminum into $9n/10$ pounds of new, virgin aluminum. How much usable aluminum will 100 pounds of virgin aluminum ultimately create, if we assume that it is continually returned to the same recycling plant?

Solution: We begin with 100 pounds of aluminum. When it is returned to the plant as scrap and recycled, $0.9 \cdot 100 = 90$ pounds of new aluminum results. When that aluminum is returned to the plant as scrap and recycled, $0.9 \cdot 90 = 81$ pounds of new aluminum results. This process continues forever. The total amount of aluminum created is therefore equal to

$$100 + 90 + 81 + \cdots$$
$$= 100 \cdot 1 + 100 \cdot 0.9 + 100 \cdot 0.9 \cdot 0.9 + \cdots$$
$$= 100(1 + 0.9 + 0.9^2 + \cdots)$$
$$= 100 \cdot \sum_{j=0}^{\infty} (0.9)^j$$
$$= 100 \cdot \frac{1}{1 - 0.9}$$
$$= 1000$$

Therefore 1000 pounds of usable aluminum is ultimately generated by an initial 100 pounds of virgin aluminum.   □

13.10 Convergence of *p*-Series

We frequently encounter series of the form

$$\sum_{j=1}^{\infty} \frac{1}{j^p}$$

for some fixed exponent p. Examples are

$$\sum_{j=1}^{\infty} \frac{1}{j^2} \quad \text{and} \quad \sum_{j=1}^{\infty} \frac{1}{j} \quad \text{and} \quad \sum_{j=1}^{\infty} \frac{1}{j^{1/2}}$$

The second of these series is the harmonic series, and is known to diverge (see Sec. 13.3). We shall learn that the first series converges and the third diverges.

A useful tool for studying series of this kind is the *Cauchy condensation test*, which we formulate as follows:

Theorem 13.4 *Let $a_1 \geq a_2 \geq a_3 \geq \cdots$ be positive numbers that tend to 0. The series*

$$\sum_{j=1}^{\infty} a_j \tag{13.9}$$

converges if and only if the series

$$\sum_{j=1}^{\infty} 2^j a_{2^j} \tag{13.10}$$

converges.

This result is best understood by examining two pictures.

In Fig. 13.2, we see that each box corresponds to (half of) a term of the condensed series (13.10). For instance, the first box has height a_1 and width $1 = 2^0$; the second box has height a_2 and width $2 = 2^1$; the third box has height a_4 and width $4 = 2^2$; and so forth. Thus any partial sum for the original series (13.9) is dominated by

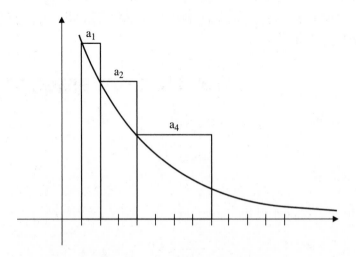

Figure 13.2 The condensed series dominates the original series.

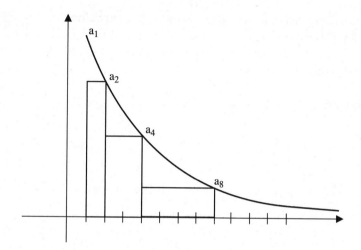

Figure 13.3   The original series dominates the condensed series.

the corresponding partial sum for the condensed series (13.10). We conclude that if the condensed series converges then the original series converges.

Now look at Fig. 13.3. Now the boxes lie *below* the graph of the $\{a_j\}$ instead of above. The first box has height a_2 and width $1 = 2^0$; the second box has height a_4 and width $2 = 2^1$; the third box has height a_8 and width $4 = 2^2$; and so forth. Now we see that half any partial sum for the condensed series (13.10) is dominated by twice the corresponding partial sum for the original series (13.9). We conclude that if the original series converges then the condensed series converges.

Thus we now have a new tool for determining the convergence of series. Let us learn from some examples.

EXAMPLE 13.28

Determine convergence for the series

$$\sum_{j=1}^{\infty} \frac{1}{j^2}$$

Solution: The summands form a decreasing sequence of positive numbers that tends to zero. So we may apply the Cauchy condensation test. We examine

$$\sum_{j=1}^{\infty} 2^j a_{2^j} = \sum_{j=1}^{\infty} 2^j \cdot \frac{1}{2^{2j}} = \sum_{j=1}^{\infty} \frac{1}{2^j}$$

Of course this is a geometric series that we have studied before. It converges, hence the original series converges.   □

EXAMPLE 13.29
Determine convergence for the series

$$\sum_{j=1}^{\infty} \frac{1}{2^{j/2}}$$

Solution: We of course apply the Cauchy condensation test. Thus we examine the series

$$\sum_{j=1}^{\infty} 2^j \frac{1}{(2^j)^{1/2}} = \sum_{j=1}^{\infty} 2^j \cdot \frac{1}{2^{j/2}} = \sum_{j=1}^{\infty} 2^{j/2}$$

This is a geometric series. But the terms do not tend to zero, so the series cannot converge. We conclude that the original series also does not converge.   □

EXAMPLE 13.30
Examine the harmonic series using the Cauchy condensation test.

Solution: The harmonic series is

$$\sum_{j=1}^{\infty} \frac{1}{j}$$

Therefore we must examine the series

$$\sum_{j=1}^{\infty} 2^j \cdot \frac{1}{2^j} = \sum_{j=1}^{\infty} 1$$

Of course this series diverges, so the harmonic series diverges (as we already know) as well.   □

In our ensuing discussions, we shall refer to a series of the form

$$\sum_{j=1}^{\infty} \frac{1}{j^p}$$

as a *p*-series.

13.11 The Comparison Test

13.11.1 A NEW CONVERGENCE TEST

Suppose that

$$\sum_{j=1}^{\infty} a_j$$

is a convergent series of nonnegative terms:

$$\sum_{j=1}^{\infty} a_j = \ell$$

Because the partial sums form an increasing sequence, these partial sums must increase to ℓ. Therefore for each partial sum S_N we may say that

$$\sum_{j=1}^{N} a_j = S_N \leq \ell$$

If

$$\sum_{j=1}^{\infty} c_j$$

is another series satisfying $0 \leq c_j \leq a_j$ for every j, then the partial sums T_N for this series satisfy

$$T_N = \sum_{j=1}^{N} c_j \leq \sum_{j=1}^{N} a_j \leq \ell$$

Thus the partial sums T_N of the series $\sum_{j=1}^{\infty} c_j$ are increasing (since the c_j's are non-negative) and bounded above by ℓ. By a property of bounded increasing sequences that we have discussed before, we conclude that the sequence of T_N's converges.

We summarize:

Theorem 13.5 *(The comparison test for convergence) Let $0 \leq c_j \leq a_j$ for every j. If the series $\sum_{j=1}^{\infty} a_j$ converges then the series $\sum_{j=1}^{\infty} c_j$ also converges.*

EXAMPLE 13.31
Show that the series

$$\sum_{j=1}^{\infty} \frac{1}{j \cdot (j^2 + 1)}$$

converges.

Solution: Observe that

$$j^2 + 1 \geq j^2$$

hence

$$j \cdot (j^2 + 1) \geq j^3$$

We conclude that

$$\frac{1}{j \cdot (j^2 + 1)} \leq \frac{1}{j^3}$$

for every j. Also

$$\sum_{j=1}^{\infty} \frac{1}{j^3}$$

is a convergent p-series. By the comparison test for convergence, the original series

$$\sum_{j=1}^{\infty} \frac{1}{j \cdot (j^2 + 1)}$$

converges.   □

EXAMPLE 13.32
Discuss convergence for the series

$$\sum_{j=1}^{\infty} \frac{1}{(3j - 2)^2}$$

Solution: Notice that

$$3j - 2 = j + (2j - 2) \geq j$$

for all $j \geq 1$. Therefore

$$(3j - 2)^2 \geq j^2$$

so that

$$\frac{1}{(3j - 2)^2} \leq \frac{1}{j^2}$$

Finally,

$$\sum_{j=1}^{\infty} \frac{1}{j^2}$$

converges by Example 13.28. By the comparison test for convergence,

$$\sum_{j=1}^{\infty} \frac{1}{(3j - 2)^2}$$

converges.   □

EXAMPLE 13.33
Discuss convergence for the series

$$\sum_{j=1}^{\infty} \frac{\sin^2(2j + 5)}{j^4 + 8j + 6} \tag{13.11}$$

Solution: Observe that

$$0 \leq \frac{\sin^2(2j + 5)}{j^4 + 8j + 6} \leq \frac{1}{j^4 + 8j + 6} \leq \frac{1}{j^4}$$

Also

$$\sum_{j=1}^{\infty} \frac{1}{j^4}$$

is a convergent *p*-series. By the comparison test for convergence, the original series (13.11) converges. □

You Try It: Discuss convergence for the series $\sum_j j/(j^3 + j)$.

EXAMPLE 13.34
Does the series

$$\sum_{j=1}^{\infty} \frac{1}{(j+2) \cdot \ln(j+2)}$$

converge?

Solution: This series has nonnegative summands and is termwise smaller than the series

$$\sum_{j=1}^{\infty} \frac{1}{j}$$

However, this last series *diverges*. So the comparison test for convergence *does not tell us anything* about the series

$$\sum_{j=1}^{\infty} \frac{1}{(j+2) \cdot \ln(j+2)}$$

Note that we are *not* saying that the series diverges. Rather, we can draw no conclusion. □

Insight: In fact the Cauchy condensation test may be applied directly to the series in the last example. First rewrite the series as

$$\sum_{j=3}^{\infty} \frac{1}{j \cdot \ln j}$$

Now apply Cauchy to obtain

$$\sum_{j=3}^{\infty} 2^j \cdot \frac{1}{2^j \ln(2^j)}$$

This simplifies to

$$\frac{1}{\ln 2} \sum_{j=3}^{\infty} \frac{1}{j}$$

Of course this is a constant multiple of the harmonic series, and it diverges. We conclude that the original series diverges as well.

Insight: Because the convergence or divergence properties of a series do not depend on the first several terms, we could have stated the comparison test for convergence by requiring that $0 \leq c_j \leq a_j$ for *all sufficiently large* j. Let us look at an example to see how this might work in practice.

EXAMPLE 13.35
Determine whether the series

$$\sum_{j=1}^{\infty} \frac{1}{j^3 - 4j^2 - 6j - 5}$$

converges.

Solution: We notice that for $j \geq 20$ we have

$$
\begin{aligned}
j^3 - 4j^2 - 6j - 5 &= j \cdot j^2 - 4j^2 - 6j - 5 \\
&\geq 20j^2 - 4j^2 - 6j - 5 \\
&= 4j^2 + (5j^2 - 4j^2) + (6j^2 - 6j) + (5j^2 - 5)
\end{aligned}
$$

Now each of the terms in parentheses is positive for $j \geq 20$. Therefore the last line exceeds $4j^2$ when $j \geq 20$. We conclude that, for large j,

$$0 \leq \frac{1}{j^3 - 4j^2 - 6j - 5} \leq \frac{1}{4j^2}$$

Because

$$\sum_{j=1}^{\infty} \frac{1}{4j^2}$$

converges, it follows from the comparison test for convergence that the series

$$\sum_{j=1}^{\infty} \frac{1}{j^3 - 4j^2 - 6j - 5}$$

converges. □

13.12 A Test for Divergence

We can now reverse our reasoning to obtain a comparison test for divergence. Namely, suppose that $0 \le c_j \le a_j$ and that the series

$$\sum_{j=1}^{\infty} c_j$$

diverges. Then the series

$$\sum_{j=1}^{\infty} a_j$$

would have to diverge also. For if the latter series converged then the comparison test for convergence would imply that

$$\sum_{j=1}^{\infty} a_j$$

converges, and that would be a contradiction. We summarize:

Theorem 13.6 *(The comparison test for divergence) Let* $0 \le c_j \le a_j$ *for every* j. *If the series* $\sum_{j=1}^{\infty} c_j$ *diverges then the series* $\sum_{j=1}^{\infty} a_j$ *also diverges.*

We note that Theorem 13.6 is, in effect, a restatement of Theorem 13.5 (in fact it is the contrapositive). But we display it as a separate theorem because of its importance.

EXAMPLE 13.36
Show that the series

$$\sum_{j=1}^{\infty} \frac{\ln(j+4)}{j} \qquad (13.12)$$

diverges.

Solution: Observe that

$$\frac{\ln(j+4)}{j} \geq \frac{1}{j}$$

for every $j \geq 1$ and the series

$$\sum_{j=1}^{\infty} \frac{1}{j}$$

diverges. By the comparison test for divergence, we conclude that the series in Eq. (13.12) diverges as well.   □

EXAMPLE 13.37
Analyze the series

$$\sum_{j=1}^{\infty} \frac{1}{\sqrt{j+2} \cdot \sqrt{\ln(j+2)}} \qquad (13.13)$$

Solution: We notice that

$$\frac{1}{\sqrt{j+2} \cdot \sqrt{\ln(j+2)}} \geq \frac{1}{(j+2) \cdot \ln(j+2)}$$

for each j. In Example 13.34 above we already noted that the series

$$\sum_{j=1}^{\infty} \frac{1}{(j+2) \cdot \ln(j+2)}$$

diverges. By the comparison theorem for divergence, the series in Eq. (13.13) diverges.   □

You Try It: Discuss convergence or divergence for the series $\sum_{j} j/(j^2 + j)$.

EXAMPLE 13.38
Use the comparison theorem for divergence to study the series

$$\sum_{j=1}^{\infty} \frac{1}{[j^{1/2} + 3]^{1/2}}$$

Solution: Observe that

$$\frac{1}{[j^{1/2} + 3]^{1/2}} \geq \frac{1}{[4j^{1/2}]^{1/2}}$$

The series

$$\sum_{j=1}^{\infty} \frac{1}{[4j^{1/2}]^{1/2}} = \frac{1}{2} \sum_{j=1}^{\infty} \frac{1}{j^{1/4}}$$

is a divergent p-series hence the series

$$\sum_{j=1}^{\infty} \frac{1}{[j^{1/2} + 3]^{1/2}}$$

also diverges.   □

EXAMPLE 13.39
Does the series

$$\sum_{j=2}^{\infty} \frac{\ln^2(j + 2)}{(j + 2)^2} \tag{13.14}$$

converge?

Solution: This series of nonnegative summands termwise exceeds the series

$$\sum_{j=1}^{\infty} \frac{1}{(j + 2)^2}$$

But the series

$$\sum_{j=1}^{\infty} \frac{1}{(j + 2)^2}$$

converges because it is a *p*-series. Thus the comparison test for divergence *does not tell us anything* about the convergence or divergence of the series

$$\sum_{j=2}^{\infty} \frac{\ln^2(j+2)}{(j+2)^2}$$ $\square$

Insight: Examples 13.38 and 13.39 remind us that the comparison tests are *not* "if and only if" statements. If the comparison test for convergence works, then the series in question converges. If the test fails, we know nothing. Likewise, if the comparison test for divergence works, then the series in question diverges. Otherwise we know nothing.

Also bear in mind that the comparison tests are valid only for series with nonnegative terms.

13.13 The Ratio Test

Theorem 13.7 *(The Ratio Test) Let*

$$\sum_{j=1}^{\infty} c_j$$

be a series. Suppose that

$$\lim_{j \to \infty} \left| \frac{c_{j+1}}{c_j} \right| = L$$

Then

   1. *If $L < 1$ then the series converges absolutely.*
   2. *If $L > 1$ then the series diverges.*
   3. *If $L = 1$ then the test gives no information.*

Insight: Take particular notice that the theorem says that when $L = 1$ then the test yields no information. This point will be made clearer in the examples. Also remember that the limit of the expression $|c_{j+1}/c_j|$ might not even exist. In this case the test also yields no information.

Finally notice that the test only depends on the *limit* of the ratio $|c_{j+1}/c_j|$. The outcome does *not* depend on the first million or so terms of the series, and that is the way it should be: the convergence or divergence of a series depends only on its "tail."

Now we look at some examples that illustrate when the ratio test gives convergence, when it gives divergence, and when it gives no information.

EXAMPLE 13.40
Apply the ratio test to the series

$$\sum_{j=1}^{\infty} \frac{2^j}{j!}$$

Solution: Recall that $j! = j \cdot (j-1) \cdot (j-2) \cdots 3 \cdot 2 \cdot 1$. Observe that $c_j = 2^j/j!$ Then

$$\left| \frac{c_{j+1}}{c_j} \right| = \frac{2^{j+1}/(j+1)!}{2^j/j!}$$

$$= \frac{2}{j+1}$$

As $j \to \infty$, the limit of this expression is 0. Therefore in this example the limit L exists and equals 0. By Part **(1)** of the ratio test, we may conclude that the series converges absolutely. ☐

EXAMPLE 13.41
Discuss the convergence or divergence of the series

$$\sum_{j=1}^{\infty} \frac{j^{10}}{2^j}$$

Solution: In this example we have $c_j = j^{10}/2^j$. Therefore

$$\left| \frac{c_{j+1}}{c_j} \right| = \frac{(j+1)^{10}/2^{j+1}}{j^{10}/2^j}$$

$$= \left(\frac{j+1}{j} \right)^{10} \cdot \frac{1}{2}$$

The limit of this last expression exists and equals 1/2. Therefore in this example $L = 1/2 < 1$. Part **(1)** of the ratio test then tells us that the series converges absolutely.   □

You Try It: Discuss convergence or divergence for the series $\sum_j (j^2 + 1)/2^j$.

EXAMPLE 13.42
Analyze the series

$$\sum_{j=1}^{\infty} \frac{(3/2)^j}{j^{3/2}}$$

Solution: We have

$$c_j = \frac{(3/2)^j}{j^{3/2}}$$

Therefore

$$\left| \frac{c_{j+1}}{c_j} \right| = \frac{(3/2)^{j+1}/(j+1)^{3/2}}{(3/2)^j/j^{3/2}}$$

$$= \left(\frac{j}{j+1} \right)^{3/2} \cdot \frac{3}{2}$$

Now, as $j \to \infty$, the last expression tends to the limit $L = 3/2$. Because $L > 1$, the ratio test says that the series diverges.   □

EXAMPLE 13.43
Apply the ratio test to the series

$$\sum_{j=1}^{\infty} \frac{1}{j}$$

Solution: We have $c_j = 1/j$ hence

$$\left| \frac{c_{j+1}}{c_j} \right| = \frac{1/(j+1)}{1/j}$$

$$= \frac{j}{j+1}$$

As $j \to \infty$, the last expression tends to 1. Therefore the ratio test gives no information whatever. Of course this is the harmonic series, and we know that it diverges.   □

EXAMPLE 13.44
Apply the ratio test to the series

$$\sum_{j=1}^{\infty} \frac{1}{j^3}.$$

Solution: Observe that $c_j = 1/j^3$. Therefore

$$\left| \frac{c_{j+1}}{c_j} \right| = \frac{1/(j+1)^3}{1/j^3} = \frac{j^3}{(j+1)^3} \to 1$$

as $j \to \infty$. Therefore the ratio test gives no information. Of course this is a p-series, and we know that it converges.   □

You Try It: Discuss convergence or divergence for the series $\sum_j 3^j/(2^j + j)$.

Insight: Examples 13.43 and 13.44 together explain what we mean when we say that the ratio test gives no information when the limit L is equal to 1. Under these circumstances, the series could diverge (Example 13.43) or the series could converge (Example 13.44). There is no way to tell: you *must use another convergence test to find out*.

13.14 The Root Test

The root test has a similar flavor to the ratio test. Sometimes it is easier to apply the ratio test than it is to apply the root test or vice versa. Thus, even though the tests look rather similar, you should be well-versed at using both of them.

Theorem 13.8 *(The root test]* Let

$$\sum_{j=1}^{\infty} c_j$$

be a series. Suppose that

$$\lim_{j \to \infty} |c_j|^{1/j} = L$$

Then

(1) *If $L < 1$ then the series converges absolutely.*
(2) *If $L > 1$ then the series diverges.*
(3) *If $L = 1$ then the test gives no information.*

Insight: Once again take particular note that when $L = 1$ then the root test gives no information whatever. Also, the limit L might not even exist. In this case, the test also gives no information. Finally, the root test does not depend on the first million or so terms of the series—only on the "tail."

Now we look at some examples that will illustrate when the root test gives convergence, when it gives divergence, and when it gives no information.

EXAMPLE 13.45
Analyze the series

$$\sum_{j=1}^{\infty} \frac{j}{(j^2 + 6)^j}$$

Solution: We apply the root test. Notice that $c_j = j/(j^2 + 6)^j$ so that

$$|c_j|^{1/j} = \frac{j^{1/j}}{j^2 + 6} \tag{13.15}$$

Certainly for any $j \geq 1$ we have that

$$\frac{j^{1/j}}{j^2 + 6} \leq \frac{j}{j^2} = \frac{1}{j} \to 0$$

It follows that the expression (13.15) tends to 0 as $j \to \infty$. Therefore $L = 0 < 1$ and the root test tells us that the series converges.   □

Insight: Notice that

$$\frac{j}{(j^2 + 6)^j} \leq \frac{j}{j^{2j}} \leq \frac{j}{j^4}$$

provided that $j \geq 2$. Thus the series from Example 13.45 is termwise dominated by the series

$$\sum_{j=1}^{\infty} \frac{1}{j^3}$$

Thus convergence can also be obtained from the comparison test. The root test, however, proves to be more straightforward to apply in this example.

EXAMPLE 13.46
Apply the root test to the series

$$\sum_{j=1}^{\infty} \frac{1}{[\ln(j+2)]^j}$$

Solution: Observe that $c_j = 1/[\ln(j+2)]^j$. Thus

$$|c_j|^{1/j} = \frac{1}{\ln(j+2)}$$

This expression tends to 0 as $j \to \infty$. Therefore $L = 0 < 1$ and the Root test says that the series converges absolutely. (Try using the ratio test on this one! How about the Comparison Test?) □

You Try It: Discuss convergence or divergence for the series $\sum_j 2^j/(4^j + 3^j)$.

EXAMPLE 13.47
Apply the root test to the series

$$\sum_{j=1}^{\infty} \frac{2^j}{j^{10}}$$

Solution: We have $c_j = 2^j/j^{10}$. Therefore

$$|c_j|^{1/j} = \frac{2}{(j^{1/j})^{10}} \tag{13.16}$$

Since $j^{1/j} \to 1$ as $j \to \infty$, the expression (13.16) tends to $2 = L > 1$. We conclude, using the root test, that the series diverges. □

EXAMPLE 13.48

Test the series

$$\sum_{j=1}^{\infty} \frac{(7j^2 + 1)^j}{(2j + 2)^{2j}}$$

for convergence or divergence.

Solution: We see that

$$c_j = \frac{(7j^2 + 1)^j}{(2j + 2)^{2j}}$$

As a result,

$$|c_j|^{1/j} = \frac{7j^2 + 1}{(2j + 2)^2}$$

$$= \frac{7j^2 + 1}{4j^2 + 8j + 4} \rightarrow \frac{7}{4} > 1$$

By the root test, the series diverges.   □

EXAMPLE 13.49

Analyze the series

$$\sum_{j=1}^{\infty} j^2$$

using the root test.

Solution: We have $c_j = j^2$. Thus

$$|c_j|^{1/j} = (j^{1/j})^2$$

This last expression tends to $L = 1$ as $j \rightarrow \infty$. Therefore the root test gives no conclusion. On the other hand, the terms of the series do not tend to zero hence the series fails the zero test. Thus the series diverges.   □

EXAMPLE 13.50

Does the series

$$\sum_{j=1}^{\infty} \frac{1}{j \cdot (\ln j)^2}$$

converge?

Solution: We see that $c_j = 1/[j \cdot (\ln j)^2]$. Therefore

$$|c_j|^{1/j} = \frac{1}{j^{1/j} \cdot [(\ln j)^{1/j}]^2}$$

As $j \to \infty$, we know that

$$j^{1/j} \to 1$$

Also

$$1 \le (\ln j)^{1/j} \le j^{1/j}$$

for $j \ge 3$ hence we may conclude that our sequence is trapped between two sequences that we have already studied. In particular, $(\ln j)^{1/j} \to 1$.

In conclusion,

$$|c_j|^{1/j} \to 1 = L$$

Therefore the root test gives absolutely no information. Check for yourself that the Cauchy condensation test gives convergence. ☐

Insight: Examples 13.49 and 13.50 explain what we mean when we say that the root test gives no information when $L = 1$. Under these circumstances the series could diverge (Example 13.49) or the series could converge (Example 13.50). *The only way to find out is to use another test.*

Exercises

In Exercises 1 and 2, write out the first ten partial sums of the series.

1. $\sum_{k=0}^{\infty} \frac{5}{2^j}$

2. $\sum_{j=1}^{\infty} 3^{-j}$

In each of Exercises 3 and 4, use a calculator to determine some partial sums and guess whether the series converges. If you can (for the cases of convergence), state what the sum should be.

3. $\sum_{j=1}^{\infty} \frac{j}{j+1}$

4. $\sum_{j=1}^{\infty} (\frac{1}{j} - \frac{1}{j+1})$

Each of the series in Exercises 5 and 6 converges. Explain why. Can you find the sum?

5. $\sum_{j=4}^{\infty} \frac{2}{(j-2)(j-3)}$

6. $\sum_{j=1}^{\infty} 2^{-j} \cos(j\pi)$

In Exercises 7 and 8, find the explicit sum of the series.

7. $\sum_{j=0}^{\infty} 8^{-j}$

8. $\sum_{j=0}^{\infty} (3/7)^j$

In Exercises 9 and 10, calculate each of the sums by using the formula for the partial sum of a geometric series which appears in the text.

9. $\sum_{j=0}^{6} 3^{-j}$

10. $\sum_{j=3}^{12} 11^{j+1}$

In Exercises 11 and 12, use the comparison tests to determine whether the series converges or diverges.

11. $\sum_{j=1}^{\infty} \frac{\sin^2 j}{j^2}$

12. $\sum_{j=1}^{\infty} \frac{j^2}{j^3+1}$

In Exercise 13 , test the given series for convergence or divergence by using the Ratio Test. If the test gives no information, then say so explicitly.

13. $\sum_{j=1}^{\infty} \frac{1}{j!}$

In Exercise 14, test the given series for convergence or divergence by using the Root Test. If the test gives no information, then say so explicitly.

14. $\sum_{j=1}^{\infty} \left(\frac{3j}{j+1}\right)^j$

Final Exam

1. What is the contrapositive of the statement "If books are good then jobs are scarce"?

   (a) If jobs are scarce then books are good.
   (b) If jobs are not scarce then books are bad.
   (c) If jobs are bad then books are scarce.
   (d) If jobs are good then books are bad.
   (e) If jobs are books then scarce is bad.

2. Give an example of a statement involving the variable x that is true for all values of x (in some universe that you will specify).

   (a) If x is a real number then $x^2 + 1 > 0$.
   (b) If x is a real number then $3 < x < 5$.
   (c) If x is a rational number then $x^3 > 0$.
   (d) If x is a real number then x is a rational number.
   (e) If x is an integer then $x > 0$.

3. Give an example of a statement involving the variable x that is false for all values of x (in some universe that you will specify).

   (a) If x is a real number then $x^4 \geq 0$.
   (b) If x is a real number then $x + 3 = 7$.
   (c) If x is an integer then $x^2 = 2$.
   (d) If x is a natural number then $x^2 < 0$.
   (e) If x is a complex number then $x^2 = x$.

4. What is the negation of the statement "All fish have fins"?

   (a) No fish have fins.
   (b) A few fish have fins.
   (c) Fins have fish.
   (d) There is a fish with no fins.
   (e) Lots of fish have no fins.

5. What is the negation of the statement "Either the woman is blonde or the man is short"?

   (a) The woman is blonde and the man is short.
   (b) The woman is not blonde and the man is tall.
   (c) The woman is tall and the man is blonde.
   (d) Everyone is blonde.
   (e) Nobody is short.

6. Give a statement that is logically equivalent to $\sim \mathbf{B} \Rightarrow \sim \mathbf{A}$ and which uses only $\vee$ and $\sim$.

   (a) $\mathbf{B} \vee \sim \mathbf{A}$
   (b) $\sim \mathbf{B} \vee \sim \mathbf{A}$
   (c) $\mathbf{B} \vee \mathbf{A}$
   (d) $\mathbf{B} \vee \sim \mathbf{B}$
   (e) $\mathbf{A} \vee \sim \mathbf{A}$

7. Give a statement that is logically equivalent to $\sim \mathbf{A} \Rightarrow \sim \mathbf{B}$ and which uses only $\wedge$ and $\sim$.

   (a) $\mathbf{B} \wedge \sim \mathbf{A}$
   (b) $\sim \mathbf{B} \wedge \sim \mathbf{A}$
   (c) $\sim (\sim \mathbf{A} \wedge \mathbf{B})$

(d) $B \wedge \sim B$

(e) $A \wedge \sim A$

8. The denial of the statement "All red boats sail on blue water" is

   (a) Some red boats sail on blue water.

   (b) There is some red boat that sails on water that is not blue.

   (c) All red boats do not sail.

   (d) Some red boats do sail.

   (e) This red boat is for sale.

9. The method of *proof by contradiction* is useful for

   (a) Proving that something is false.

   (b) Proving that something is undecidable.

   (c) Proving that something is true.

   (d) Proving that something cannot be proved.

   (e) Disproving something.

10. The method of mathematical induction involves

    (a) Infinitely many steps.

    (b) A contradictory step.

    (c) A bootstrap step.

    (d) An important inductive step.

    (e) An indeterminate step.

11. The method of direct proof is important because

    (a) It is indirect.

    (b) It is valid.

    (c) It is invalid.

    (d) It applies to most proofs.

    (e) It is logically simple.

12. If $A = \{1, 2, 3\}$ and $B = \{2, 3, 4\}$ then

    (a) $A \subset B$

    (b) $B \subset A$

    (c) $A \cap B = \emptyset$

(d) $A \cap B$ has two elements

(e) $A \cup B$ has three elements

13. If $A \subset B$ and $B \subset C$ then

   (a) $A \supset C$

   (b) $A \supset B$

   (c) $A \cap B = C$

   (d) $A \cap C = B$

   (e) $A \subset C$

14. If A has 2 elements and B has 3 elements and C has 4 elements, then $A \times B \times C$ has

   (a) 4 element

   (b) 9 elements

   (c) 15 elements

   (d) 24 elements

   (e) 6 elements

15. If $x \in A$ and $A \in B$ then

   (a) x may not be an element of B

   (b) $x \in B$

   (c) $x \cap B = \emptyset$

   (d) $x \cup B = A$

   (e) $x \cap A = B$

16. The set $^c(A \cap B)$ is the same as

   (a) $A \cap B$

   (b) $A \cup B$

   (c) $^cA \cup {}^cB$

   (d) $A \subset B$

   (e) $B \subset A$

17. If A has 5 elements and B has 4 elements then

   (a) $A \cap B$ has 3 elements.

   (b) $A \cup B$ has 6 elements.

   (c) $A \times B$ has 10 elements.

   (d) $A \cap B$ has at most 4 elements.

   (e) $A \backslash B$ has at most 2 elements.

18. Let S be the collection of all sets with at most 5 elements. Then

   (a) An element of A is a set with 1, 2, 3, 4, or 5 elements.
   (b) An element of S is a number.
   (c) An element of S is a set with 25 elements.
   (d) An element of S is a set with an arbitrary number of elements.
   (e) An element of S is a superset of S.

19. If f is a function with domain S and values in $\mathbb{R}$ and g is a function with domain S and values in $\mathbb{R}$ then

   (a) $f - g$ is a function with domain S and values in $\mathbb{C}$.
   (b) $f \cdot g$ is undefined.
   (c) All the arithmetic operations make sense on f and g (provided we do not divide by 0).
   (d) There are no functions between f and g.
   (e) f is smaller than g.

20. Say that two real numbers a and b are related if the sum of their squares is 100. Then this relation is

   (a) A function.
   (b) One with finitely many elements.
   (c) An equivalence relation.
   (d) Symmetric in its entries.
   (e) Increasing.

21. Say that two functions f and g, with domain $\mathbb{R}$, are related if $f(x) \le g(x)$ for every $x \in \mathbb{R}$. Then

   (a) This is an equivalence relation.
   (b) This is an order relation.
   (c) This is a function.
   (d) This is a total ordering.
   (e) This does not allow us to compare any two functions.

22. If f is a natural-number-valued function with domain $\mathbb{Q}$ and g is a natural-number-valued function with domain $\mathbb{Q}$ then $h(x) = f(x)^{g(x)}$ is

   (a) A function with domain $\mathbb{Q}$ and image $\mathbb{C}$.
   (b) A function with values in $\mathbb{N}$.
   (c) A function with upper and lower bounds.

(d) A function with limited values.

(e) A function with bounded socles.

23. A function $f(x)$ with domain and range $\mathbb{R}$ is said to be the *square root* of the function g with domain and range $\mathbb{R}$ if $f^2(x) = g(x)$ for every x. If f and g have this relationship then

(a) The function f is greater than the function g.

(b) The function g takes nonnegative values.

(c) The function f is increasing.

(d) The function g is differentiable.

(e) The function f is exceptional.

24. If $f(x) = x^2$ and $g(x) = \sin x$ then

(a) $f \circ g$ is unbounded.

(b) $g \circ f$ is unbounded.

(c) $f \circ g$ and $g \circ f$ are both well defined.

(d) $f \cdot g$ is increasing.

(e) $f - g$ is monotone.

25. Say that two universities are related if they sponsor NCAA sports and if they play against each other regularly. Then

(a) This is a symmetric relation.

(b) This is a doomed relation.

(c) This is a function.

(d) This is an equivalence relation.

(e) This is a perfect relation.

26. The union of two functions is

(a) Always an equivalence relation.

(b) Never an equivalence relation.

(c) Another function.

(d) A one-to-one function.

(e) An onto function.

27. Between every two distinct rational numbers there is

(a) A whole number or integer.

(b) A multiple of five.

(c) An irrational number.

    (d) A quintessential number.

    (e) A complex number.

28. Addition of integers is

    (a) Commutative and associative.

    (b) Distributive.

    (c) Disruptive.

    (d) Meaningful.

    (e) Anticommutative.

29. The familiar number systems which form a field are

    (a) The integers and the natural numbers.

    (b) Only the rational numbers.

    (c) Only the real numbers.

    (d) The rational numbers, the real numbers, and the complex numbers.

    (e) Only the integers.

30. Every nonzero complex number has

    (a) A square root.

    (b) A cube root.

    (c) Three square roots.

    (d) No real roots.

    (e) Two distinct square roots.

31. Both square roots of a nonzero, positive real number will be

    (a) Real.

    (b) Positive.

    (c) Negative.

    (d) Complex.

    (e) Incommensurable.

32. Unlike the rational numbers, the real number system is

    (a) Unbounded.

    (b) Complete.

    (c) Monotone.

(d) Decreasing.

(e) Decentralized.

33. The process of mathematical induction is closely related to

(a) The rational number system.

(b) The integers.

(c) The natural numbers.

(d) The real numbers.

(e) The quaternions.

34. The complex numbers improve or augment the reals in that

(a) They are more complex.

(b) Every polynomial has a root.

(c) They are two-dimensional rather than one-dimensional.

(d) They have Argand diagrams.

(e) They are commutative under addition.

35. The quaternions do not form a field because

(a) They are four-dimensional.

(b) They have **i**, **j**, and **k** as elements.

(c) They are commutative under addition.

(d) They are not commutative under multiplication.

(e) They are more complex than the complex numbers.

36. The modulus of a complex number measures

(a) Its distance from the origin.

(b) Its distance from the real axis.

(c) Its distance from the imaginary axis.

(d) Its size and shape.

(e) Its magnitude.

37. The triangle inequality is

(a) A special property of the real number system.

(b) A special property of the complex number system.

(c) A special property of the rational number system.

(d) A useful device for measuring distance in a number system.

(e) A fact about triangles.

38. The rational number system is

(a) Closed under limits.

(b) A very small number system.

(c) Not closed under square roots.

(d) Almost as big as the real number system.

(e) Much smaller than the complex number system.

39. The existence of negative integers is

(a) An artifice that we concoct to make life interesting.

(b) A property that we hope is true.

(c) A special feature of that number system.

(d) A byproduct of the way that we construct the integers.

(e) Very confusing.

40. In everyday commerce and in most practical activities the number systems that are most commonly used are

(a) The quaternions.

(b) The integers and the rational numbers.

(c) The natural numbers.

(d) The complex numbers.

(e) The algebraic integers.

41. The number of different ways to select 3 cards from a pack of 10 is

(a) 120

(b) 100

(c) 50

(d) 96

(e) 84

42. The number of distinct permuations of 6 objects is

(a) 500

(b) 720

(c) 650

(d) 401

(e) 405

43. Fifty black balls and 50 white balls are dropped at random into 60 cups. We can be sure that

    (a) Some cup contains both a black ball and a white ball.

    (b) Some cup contains 2 black balls.

    (c) Some cup contains 2 white balls.

    (d) Some cup contains 2 balls.

    (e) Some cup contains 3 balls.

44. The expression $[x + a]^4$ can be expanded to

    (a) $x^2 + 2ax + a^2$

    (b) $x^3 + 3a^2x + 3ax^2 + a^3$

    (c) $x^4 + 4x^3a + 6x^2a^2 + 4xa^3 + a^4$

    (d) $x^4 + a^4$

    (e) $x^4 \cdot a^4$

45. The recursion relation $a_0 = 1$, $a_1 = 2$, $a_j = 2a_{j-1} - a_{j-2}$ has the solution

    (a) $a_j = j^2 + 4$

    (b) $a_j = 3j - 5$

    (c) $a_j = j^2 - j$

    (d) $a_j = j$

    (e) $a_j = 1$

46. The probability of getting a straight (5 cards in sequence) from an ordinary 52-card deck of playing cards is

    (a) 0.000189

    (b) 0.00111

    (c) 0.0123

    (d) 0.5432

    (e) 0.0000765

47. The probability of picking 2 cards at random from a standard 52-card deck and having them both be of the same denomination (in other words, the probability of having a pair) is

    (a) 0.0588

    (b) 0.0033

(c) 0.00004

(d) 0.0294

(e) 0.1111

48. You have a deck of five distinct cards, numbered 1, 2, 3, 4, and 5. You close your eyes and pick them one by one—in some random order. What is the probability that you chose them in the order 1–2–3–4–5?

(a) 0.1234

(b) 0.0044

(c) 0.00833

(d) 0.001101

(e) 0.11223

49. The sum of the matrices

$$\begin{pmatrix} 3 & 5 \\ 4 & -6 \end{pmatrix} \quad \text{and} \quad \begin{pmatrix} 1 & 9 \\ 3 & -2 \end{pmatrix}$$

is

(a) $\begin{pmatrix} 3 & 3 \\ 2 & 2 \end{pmatrix}$

(b) $\begin{pmatrix} 4 & 5 \\ 6 & 7 \end{pmatrix}$

(c) $\begin{pmatrix} 4 & 14 \\ 7 & -8 \end{pmatrix}$

(d) $\begin{pmatrix} 1 & 1 \\ 1 & 1 \end{pmatrix}$

(e) $\begin{pmatrix} -2 & -5 \\ -8 & -1 \end{pmatrix}$

50. The matrix product

$$\begin{pmatrix} 3 & 3 \\ 2 & 2 \end{pmatrix} \cdot \begin{pmatrix} 1 & -2 \\ 5 & 0 \end{pmatrix}$$

equals

(a) $\begin{pmatrix} 3 & 1 \\ 1 & 2 \end{pmatrix}$

(b) $\begin{pmatrix} 18 & -6 \\ 12 & -4 \end{pmatrix}$

(c) $\begin{pmatrix} 10 & -3 \\ 2 & 6 \end{pmatrix}$

(d) $\begin{pmatrix} 1 & 1 \\ 0 & 0 \end{pmatrix}$

(e) $\begin{pmatrix} 11 & 12 \\ 13 & 14 \end{pmatrix}$

51. The inverse of the matrix $\begin{pmatrix} 0 & 1 \\ 1 & 1 \end{pmatrix}$ is

(a) $\begin{pmatrix} 1 & 1 \\ 1 & 1 \end{pmatrix}$

(b) $\begin{pmatrix} -1 & 0 \\ -1 & 1 \end{pmatrix}$

(c) $\begin{pmatrix} 2 & 1 \\ 1 & 2 \end{pmatrix}$

(d) $\begin{pmatrix} 1 & 0 \\ 0 & 1 \end{pmatrix}$

(e) $\begin{pmatrix} -1 & 1 \\ 1 & 0 \end{pmatrix}$

52. The solution of the system

$$x - 2y + z = 4$$
$$2x + y - z = 1$$
$$-x + y + z = 3$$

is

(a) $x = 2, y = 4, z = 3$
(b) $x = 2, y = 1, z = 4$

(c) $x = 5, y = 3, z = 1$

(d) $x = 1, y = 1, z = 0$

(e) $x = -3, y = -1, z = 1$

53. When we solve four equations in three unknowns we generically expect

   (a) No solutions.

   (b) One solution.

   (c) Three solutions.

   (d) Infinitely many solutions.

   (e) One solution for each variable.

54. Find the extrema of the linear function $f(x, y) = 3x - 4y$ on the planar region $\{(x, y) : 4 \le x + y \le 7, 1 \le x \le 3\}$.

   (a) The minimum value is -8 and the maximum value is 5.

   (b) The minimum value is -2 and the maximum value is 6.

   (c) The minimum value is -21 and the maximum value is 5.

   (d) The minimum value is -2 and the maximum value is 7.

   (e) The minimum value is -1 and the maximum value is 1.

55. The reason that the kth row of Pascal's triangle sums to 2^k is that

   (a) The number 2^k is even.

   (b) There are 2^k rows in Pascal's triangle.

   (c) Power of 2 are powerful.

   (d) The entries are the binomial coefficients.

   (e) There is hidden symmetry.

56. A graph without an Euler path is

   (a) Very simple.

   (b) Very complicated.

   (c) Very reducible.

   (d) Redundant.

   (e) Superfluous.

57. A complete graph on 6 vertices has

   (a) 12 edges.

   (b) 10 edges.

(c) 15 edges.

    (d) 8 edges.

    (e) 4 edges.

58. The number of different graphs on 3 vertices is

    (a) 4

    (b) 10

    (c) Infinite

    (d) Few

    (e) Many

59. A graph with four vertices and two edges

    (a) Might be connected or might not.

    (b) Must be connected.

    (c) Cannot have any Euler paths.

    (d) Must have Euler paths.

    (e) Must be reducible.

60. The Euler characteristic of a sphere with g handles is

    (a) $3g$

    (b) $2 - 2g$

    (c) $2 + 2g$

    (d) g

    (e) $4 + g$

61. A traveling salesman must visit three cities—each just once. There are paths connecting every city to every other city. He/she will begin at a particular city A. How many different routes could he/she take?

    (a) 5

    (b) 3

    (c) 2

    (d) 4

    (e) 1

62. Answer the question in Prob. 61 for four cities.

    (a) 2

    (b) 4

(c) 3

   (d) 5

   (e) 1

63. Simplify the expression $3 + 4 \bmod 5$.

   (a) 1

   (b) 2

   (c) 3

   (d) 4

   (e) 5

64. The expression $8/5 \bmod 11$ simplifies to

   (a) 6

   (b) 5

   (c) 4

   (d) 3

   (e) 2

65. Simplify the expression $6 \cdot 7 \bmod 9$.

   (a) 1

   (b) 2

   (c) 4

   (d) 5

   (e) 6

66. The identity element in a group is

   (a) One of a finite set.

   (b) Part of the reproducing set.

   (c) Unique.

   (d) Self-reproducing.

   (e) Permanent.

67. If g is an element of the group G then its multiplicative inverse element is

   (a) Unique.

   (b) Self-replicating.

   (c) Permanent.

(d) Multiplicative.

(e) Subjunctive.

68. The group $\mathbb{Z}/4\mathbb{Z}$ has how many subgroups?

    (a) 5

    (b) 3

    (c) 4

    (d) 1

    (e) 2

69. How many Sylow subgroups does $\mathbb{Z}/4\mathbb{Z}$ have?

    (a) 1

    (b) 3

    (c) 4

    (d) 2

    (e) 5

70. Can $\mathbb{Z}/3\mathbb{Z}$ be realized as a subgroup of $\mathbb{Z}/9\mathbb{Z}$ (that is, is there a subgroup that is equivalent or isomorphic to it?)?

    (a) Yes.

    (b) No.

    (c) Sometimes.

    (d) Usually.

    (e) It is forbidden.

71. Let G be a group and $g, h \in G$. Then $(g^2 h)^{-1}$ equals

    (a) $h^2 g$

    (b) $h^{-1} g^{-1}$

    (c) $h^{-1} g^{-2}$

    (d) $h^{-1} h^{-2}$

    (e) $g^{-2} h^{-1}$

72. Let G be a group and H a subgroup. Define G/H to be the set of equivalence classes of G under the relation $g \sim h$ if $g^{-1} h \in H$. Describe G/H in case $G = \mathbb{Z}$ and $H = 3\mathbb{Z}$.

    (a) $\mathbb{Z}/2\mathbb{Z}$

    (b) $\mathbb{Z}/3\mathbb{Z}$

(c) $\mathbb{Z}/4\mathbb{Z}$

(d) $\mathbb{R}$

(e) $\mathbb{Q}$

73. Call H a *normal subgroup* of the group G if $g^{-1}hg \in H$ whenever $g \in G$ and $h \in H$. Is the set of even integers a normal subgroup of $\mathbb{Z}$?

   (a) No.

   (b) Not usually.

   (c) Sometimes.

   (d) Yes.

   (e) Disallowed.

74. Refer to Prob. 73 for the definition of normal subgroup. Is The set of diagonal matrices a subgroup of the set of all 2×2 invertible matrices under multiplication?

   (a) Sometimes.

   (b) Most times.

   (c) Seldom times.

   (d) No.

   (e) Yes.

75. The encryption of the message COUNT TO FOUR under the affine encryption $P \mapsto 4P - 2$ is

   (a) HPQRSUTMNOM

   (b) GEAAYYESEAQ

   (c) ORMKBOESPKS

   (d) VFUODBNEUKD

   (e) XBGKDUNGEUM

76. The encryption of the message USE YOUR HEAD under the encryption $P \mapsto P^2 + P$ is

   (a) ZMFNETHELSEL

   (b) FLEJBIPALBME

   (c) VDEUBGOIKSUE

   (d) DDUECEUEUAUQ

   (e) FELDJBOIWJBE

77. The encrypted message KFLMUDU can be decrypted with the mapping $P \mapsto 3P + 5$. The message reads

    (a) JUMP NOW

    (b) GET HOME

    (c) STAY PUT

    (d) BACK OFF

    (e) GET GOOD

78. The digraph method breaks a message up into

    (a) Units of 3 characters.

    (b) Units of 10 characters.

    (c) Units of 2 characters.

    (d) Units of 4 characters.

    (e) Units of 7 characters.

79. When using the digraph method, we do arithmetic modulo

    (a) 676

    (b) 625

    (c) 601

    (d) 699

    (e) 666

80. Encryption and decryption are

    (a) Complementary processes.

    (b) Evil processes.

    (c) Secret processes.

    (d) Inverse processes.

    (e) Subtle processes.

81. Cryptography is an old idea, going back even to

    (a) Hannibal

    (b) Attila the Hun

    (c) William the Conqueror

    (d) Julius Caesar

    (e) Abraham Lincoln

82. The boolean expression $[a \times \overline{b}] + [\overline{a} \times b] + [a \times b]$ simplifies to

    (a) $a \times b$
    (b) $\overline{a} \times b$
    (c) $a + b$
    (d) $a + \overline{b}$
    (e) $\overline{a} \times \overline{b}$

83. Boolean algebra is useful in

    (a) Square dancing.
    (b) Circuit design.
    (c) Mosaic tiling.
    (d) Fly fishing.
    (e) Acrobatics.

84. How many axioms does boolean algebra have?

    (a) Two
    (b) Three
    (c) Five
    (d) Seven
    (e) Nine

85. The sequence $\dfrac{j^2}{j^4 + 1}$

    (a) Converges.
    (b) Diverges.
    (c) Oscillates.
    (d) Fiddles around.
    (e) Dies.

86. The sequence $j \cdot \sin j$

    (a) Converges.
    (b) Diverges.
    (c) Perpetrates.
    (d) Disintegrates.
    (e) Propagates.

87. The sequence j

   (a) Subverges.
   (b) Converges.
   (c) Diverges to ∞.
   (d) Disverges.
   (e) Postverges.

88. The sum of two sequences is

   (a) A series.
   (b) A product.
   (c) A panoply.
   (d) Another sequence.
   (e) A group.

89. If $\{a_j\}$ is a sequence of nonvanishing terms that converges to some nonzero number ℓ, then the sequence $\dfrac{1}{a_j}$

   (a) Converges to $\dfrac{1}{\ell}$.
   (b) Diverges.
   (c) Disbands.
   (d) Converges to ℓ^2.
   (e) Converges to 3ℓ.

90. The sequence $a_j = (-2)^j$

   (a) Converges.
   (b) Diverges.
   (c) Mutates.
   (d) Obfuscates.
   (e) Rotates.

91. The sequence $\dfrac{j^2}{2^j}$

   (a) Increases.
   (b) Diverges.
   (c) Converges.

(d) Displays.

(e) Radiates.

92. The pinching theorem is a device for

    (a) Proving the convergence of a sequence.

    (b) Proving the divergence of a sequence.

    (c) Recognizing a sequence.

    (d) Discarding a sequence.

    (e) Hiding a sequence.

93. The series $\sum_{j} \dfrac{j^2}{2^j}$

    (a) Diverges.

    (b) Disbands.

    (c) Refutes.

    (d) Converges.

    (e) Dissipates.

94. The series $\sum_{j} \left(\dfrac{1}{3j} \right)^{j}$

    (a) Converges.

    (b) Diverges.

    (c) Subverts.

    (d) Reverts.

    (e) Exerts.

95. The series $\sum_{j} j \cdot \sin j$

    (a) Converges.

    (b) Diverges.

    (c) Implodes.

    (d) Divests.

    (e) Soars.

96. The series $\sum_{j} \left(1 + \dfrac{1}{j} \right)^{j}$

(a) Redounds.

    (b) Exerts.

    (c) Restarts.

    (d) Converges.

    (e) Diverges.

97. The series $\displaystyle\sum_{j}\left(\frac{1}{j}\right)^{j}$

    (a) Projects.

    (b) Converges.

    (c) Rejects.

    (d) Diverges.

    (e) Subjects.

98. The sum of two series

    (a) Is one spicy meatball.

    (b) Is opinionated.

    (c) Is another series.

    (d) Is eternal.

    (e) Is ephemeral.

99. The purpose of series is to provide

    (a) A generalization of ordinary addition.

    (b) A good use of time.

    (c) A good use of money.

    (d) A diversion.

    (e) An engagement.

100. TRUE OR FALSE: If $a_j > 0$ and $\sum_j a_j$ converges then $\sum_j a_j^2$ converges.

    (a) True

    (b) False

Solutions

1. (b)	21. (e)	41. (a)	61. (c)	81. (d)
2. (a)	22. (b)	42. (b)	62. (b)	82. (c)
3. (d)	23. (b)	43. (d)	63. (b)	83. (b)
4. (d)	24. (c)	44. (c)	64. (a)	84. (e)
5. (e)	25. (d)	45. (d)	65. (e)	85. (a)
6. (a)	26. (c)	46. (a)	66. (c)	86. (b)
7. (c)	27. (c)	47. (a)	67. (a)	87. (c)
8. (b)	28. (a)	48. (d)	68. (b)	88. (d)
9. (c)	29. (d)	49. (c)	69. (d)	89. (a)
10. (d)	30. (e)	50. (b)	70. (a)	90. (b)
11. (e)	31. (a)	51. (e)	71. (c)	91. (c)
12. (d)	32. (b)	52. (b)	72. (b)	92. (a)
13. (e)	33. (c)	53. (a)	73. (d)	93. (d)
14. (d)	34. (b)	54. (c)	74. (e)	94. (a)
15. (a)	35. (d)	55. (d)	75. (b)	95. (b)
16. (c)	36. (a)	56. (d)	76. (d)	96. (e)
17. (d)	37. (d)	57. (c)	77. (a)	97. (b)
18. (a)	38. (c)	58. (c)	78. (c)	98. (c)
19. (c)	39. (x)	59. (a)	79. (a)	99. (a)
20. (d)	40. (b)	60. (b)	80. (d)	100. (a)

Solutions to Exercises

This book has a great many exercises. For some we provide sketches of solutions and for others we provide just the answers. For some, where there is repetition, we provide no answer. For the sake of mastery, we encourage the reader to *write out complete solutions* to all problems.

Chapter 1

	S	T	S ∧ T	S ∨ T	~ (S ∨ T)	(S ∧ T)∨ ~ (S ∨ T)
	T	T	T	T	F	T
1. (a)	T	F	F	T	F	F
	F	T	F	T	F	F
	F	F	F	F	T	T

S	T	S ∨ T	S ∧ T	(S ∨ T) ⇒ (S ∧ T)
T	T	T	T	T
T	F	T	F	F
F	T	T	F	F
F	F	F	F	T

(b)

2. (a) $\mathbf{S} \Rightarrow \sim \mathbf{T}$

   (b) $U \Rightarrow (V \vee \sim S)$

3. (a) If either all politicians are honest or no men are fools then I do not have two brain cells to rub together.

   (b) Either the pie is in the sky or both some men are fools and I do have two brain cells to rub together.

4. (a) **Converse:** If there are clouds, then it will rain.
   Contrapositive: If there are no clouds, then it will not rain.

   (b) **Converse:** If it is raining, then there are clouds.
   Contrapositive: If it is not raining, then there are no clouds.

5. (a) False. The area inside a circle is πr^2.

   (b) True. We note that $2 + 2 = 4$ is true and 2/5 is also a rational number.

6. (a) $\sim \mathbf{S} \vee \sim \mathbf{T}$

   (b) $\sim (\mathbf{S} \vee \mathbf{T})$

7. (a) The set S contains at most one integer.

   (b) Either some mare does not eat oats or some doe does not eat oats.

A	B	∼ A	∼ B	A ∨ ∼ B	∼ A ⇒ B
T	T	F	F	T	T
T	F	F	T	T	T
F	T	T	F	F	T
F	F	T	T	T	F

8. (a)

The two statements are logically inequivalent.

A	B	∼ A	∼ B	A ∧ ∼ B	∼ A ⇒ ∼ B
T	T	F	F	F	T
T	F	F	T	T	T
F	T	T	F	F	F
F	F	T	T	F	T

(b)

The two statements are logically inequivalent.

Chapter 2

1. If $m = 2r + 1$ and $n = 2s + 1$ then $m \cdot n = (2r + 1) \cdot (2n + 1) = 4rs + 2r + 2s + 1 = 2(2rs + r + s) + 1$, which is odd.

2. If $n = 2r$ then $m \cdot n = m \cdot (2r) = 2(m \cdot n)$, which is even.

3. By induction:
   The case $n = 1$ is true because

$$\frac{2 \cdot 1^3 + 3 \cdot 1^2 + 1}{6} = 1 = 1^2$$

Now assume that the case n has been established, so

$$1^2 + 2^2 + \cdots + n^2 = \frac{2n^3 + 3n^2 + n}{6}$$

Add $(n + 1)^2$ to both sides to obtain

$$1^2 + 2^2 + \cdots + n^2 + (n + 1)^2 = \frac{2n^3 + 3n^2 + n}{6} + (n + 1)^2$$

We may write the right-hand side as

$$\frac{2n^3 + 3n^2 + n}{6} + (n^2 + 2n + 1) = \frac{2n^3 + 9n^2 + 13n + 6}{6}$$
$$= \frac{2(n + 1)^3 + 3(n + 1)^2 + (n + 1)}{6}$$

That completes the inductive step.

4. If $m = 3^k$ and $n = 3^\ell$ and $k < \ell$ then

$$m + n = 3^k + 3^\ell = 3^k(1 + 3^{\ell-k})$$

Clearly $1 + 3^{\ell-k}$ is not a power of 3.

5. Say that $n = (a/b)^2$, where a and b have no common divisors. Then

$$nb^2 = a^2$$

Now each prime divisor of a must divide the left-hand side. But it cannot divide b, so it must divide n. And in fact it must do so twice (since a is squared on the right). So $n = a^2 r$ for some integer r. Thus

$$(a^2 r)b^2 = a^2$$

Dividing out a^2 gives

$$r b^2 = 1$$

We conclude that $r = 1$ and $b = 1$ (since all numbers here are integers). But then a/b is a whole number, not a fraction.

6. If $n + 1 = b^2$ and $n = a^2$ then

$$1 = (n + 1) - n = b^2 - a^2 = (b - a)(b + a)$$

Since all numbers are integers we must conclude that

$$b + a = 1$$
$$b - a = 1$$

Thus $a = 0$ and $b = 1$. That is a contradiction because 0 is not a natural number.

7. The inductive step is not well defined. It is not possible to write down an inductive statement that is valid for the entire argument.

8. The case $k = 3$ is trivial since $2^3 > 1 + 2 \cdot 3$. Assuming that we have established the case k, we have

$$2^k > 1 + 2k$$

Muliplying both sides by 2 we find that

$$2^{k+1} > 2 + 4k$$
$$= 2(k + 1) + 2k$$
$$> 2(k + 1) + 1$$

That is the inductive step.

9. The pigeonhole principle is clear for 1 mailbox and 2 letters.

Suppose it has been established for k mailboxes. We now have $k + 1$ mailboxes and $k + 2$ letters. If all the letters are placed in just k mailboxes then the result is clear by the inductive step. So suppose instead that every box contains at least one letter. If the first k boxes each have precisely one letter, then the last box has two and we are done. If instead one of the first k boxes has two letters then we are done. That completes the inductive step.

10. If one letter is in the wrong envelope then two letters are in the wrong envelope. So the probability is 0.

Chapter 3

1. (a) $S \cap U = \{1, 2, 3, 4\}$

   (b) $(S \cap T) \cup U = \{1, 2, 3, 4, 5, 9\}$

2. $S \times T = \emptyset$

3. (a) If $x \in S \cap (T \cup U)$ then $x \in S$ and $x \in T \cup U$. Thus $x \in S$ and either $x \in T$ or $x \in U$. So $x \in S \cap T$ or $x \in S \cap U$. Thus $x \in (S \cap T) \cup (S \cap U)$. So $S \cap (T \cup U) \subset (S \cap T) \cup (S \cap U)$.

   If now $x \in (S \cap T) \cup (S \cap U)$ then $x \in S \cap T$ and $x \in X \cap U$. So either x is in both S and T *or* x is in both S and U. Thus $x \in S$ and either $x \in T$ or $x \in U$. In conclusion, $x \in S$ and $x \in T \cup U$. We see then that $x \in S \cap (T \cup U)$. So $(S \cap T) \cup (S \cap U) \subset S \cap (T \cup U)$.

   (b) Similar.

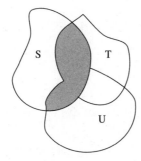

Figure 3.1s $S \cap (T \cup U) = (S \cap T) \cup (S \cap U)$.

4. (a)

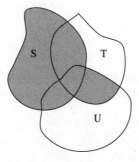

Figure 3.2s $S \cup (T \cap U) = (S \cup T) \cap (S \cup U)$.

   (b)

5. We see that $A \setminus B = \emptyset$, $A \setminus C = \emptyset$, and $A \cup B = B$.

6. We see that $\mathbb{Q} \setminus \mathbb{Z}$ equals all rational numbers, expressed in lowest terms, with denominator unequal to 1.

7. The set $\mathbb{Q} \times \mathbb{R}$ is the set of ordered pairs such that the first entry is a rational number and the second entry is a real number. The set $\mathbb{Q} \times \mathbb{Z}$ is the set of ordered pairs such that the first entry is a rational number and the second entry is an integer.

8. The set $(\mathbb{Q} \times \mathbb{R}) \setminus (\mathbb{Z} \times \mathbb{Q})$ is the set of ordered pairs such that the first entry is a rational number and the second entry is a real number, and further so that not both the first entry is an integer and the second entry is rational.

9. The power set is

$$\{\emptyset, \{a\}, \{b\}, \{1\}, \{2\}, \{a, b\}, \{a, 1\}, \{a, 2\}, \{b, 1\}, \{b, 2\}, \{1, 2\},$$

$$\{a, b, 1\}, \{a, b, 2\}, \{a, 1, 2\}, \{b, 1, 2\}, \{a, b, 1, 2\}\}$$

10. (a) False

(b) True

(c) False

11. (a) The power set is

$$\{\emptyset, \{1\}, \{\emptyset\}, \{\{a, b\}\}, \{1, \emptyset\}, \{1, \{a, b\}\}, \{\emptyset, \{a, b\}\}, \{1, \emptyset, \{a, b\}\}\}$$

(b) The power set is

$$\{\emptyset, \{\bullet\}, \{\triangle\}, \{\partial\}, \{\bullet, \triangle\}, \{\bullet, \partial\}, \{\triangle, \partial\}, \{\bullet, \triangle, \partial\}\}$$

Chapter 4

1. **Reflexive:** If $n \in \mathbb{Z}$, then $n + n = 2n$ is even so $(n, n) \in \mathcal{R}$.
   Symmetric: If $(m, n) \in \mathcal{R}$, then $m + n$ is even so $n + m$ is even hence $(n, m) \in \mathcal{R}$.
   Transitive: If $(m, n) \in \mathcal{R}$ and $(n, p) \in \mathcal{R}$, then $m + n = 2r$ is even and $n + p = 2s$ is even. Thus

$$m + n + n + p = 2r + 2s$$

and therefore

$$m + p = 2r + 2s - 2n = 2(r + s - n)$$

We conclude that $m + p$ is even so that $(m, p) \in \mathcal{R}$.

The equivalence classes are the set of even integers and the set of odd integers.

2. **Reflexive:** If $(m, n) \in \mathbb{Z} \times (\mathbb{Z} \backslash \{0\})$, then $m \cdot n = m \cdot n$ so that (m, n) $\mathcal{R}(m, n)$.

   Symmetric: If $(m, n)\mathcal{R}(m', n')$ then $m \cdot n' = m' \cdot n$ so that $m' \cdot n = m \cdot n'$. Hence $(m', n')\mathcal{R}(m, n)$.

   Transitive: If $(m, n)\mathcal{R}(m', n')$ and $(m', n')\mathcal{R}(m'', n'')$, then

$$m \cdot n' = m' \cdot n \qquad \text{and} \qquad m' \cdot n'' = m'' \cdot n'$$

   Hence

$$m \cdot n' \cdot m' \cdot n'' = m' \cdot n \cdot m'' \cdot n'$$

   Cancelling $m' \cdot n'$ from both sides, we find that

$$m \cdot n'' = m'' \cdot n$$

   Hence $(m, n)\mathcal{R}(m'', n'')$.

   The equivalence classes are ordered pairs (m, n) such that the ratio of m to n represent the same fraction. For instance, $(1, 2)$, $(3, 6)$, and $(10, 20)$ are in the same equivalence class.

3. **Reflexive:** If $(x, y) \in \mathbb{R}^2$, then $y = y$ so $(x, y)\mathcal{R}(x, y)$.

   Symmetric: If $(x, y)\mathcal{R}(x', y')$ then $y = y'$ so that $y' = y$ hence $(x', y')\mathcal{R}(x, y)$.

   Transitive: If $(x, y)\mathcal{R}(x', y')$ and $(x', y')\mathcal{R}(x'', y'')$ then $y = y'$ and $y' = y''$ so that $y = y''$. Hence $(x, y)\mathcal{R}(x'', y'')$.

   The equivalence classes are horizontal lines. A useful representative for each equivalence class is the point where the horizontal line crosses the y-axis.

4. **Reflexive:** If a is a person then a is the same as a so $a\mathcal{R}a$.

   Symmetric: If $a\mathcal{R}b$ then a and b are siblings (or the same person) with the same parents. Hence b and a are siblings (or the same person) with the same parents. We conclude that $b\mathcal{R}a$.

   Transitive: If $a\mathcal{R}b$ and $b\mathcal{R}c$ then a and b are siblings (or the same person) with the same parents and b and c are siblings (or the same person) with the same parents. Thus a and c are siblings (or the same person) with the same parents.

The equivalence classes are sets of siblings in the same family with the same parents.

5. (a) Function

   (b) Not a function

6. (a) One-to-one

   (b) Not one-to-one

7. (c)

$$f \cdot g = \{(x, t \cdot t') : (s, t) \in f, (s, t') \in g\}$$

Parts (b) and (d) are similar.

8. (a) Domain $= \{x \in \mathbb{R} : x \geq 0\}$
   Image $= \{y \in \mathbb{R} : y \geq -3\}$

   (b) Domain $=$ all people
   Image $=$ all male parents

9. Solution omitted.

10. First show that $c < a + b$. Write $c = a + b - \gamma$. Then examine $a^2 + b^2 = (a + b - \gamma)^2$.

Chapter 5

1. The number system $\mathbb{Q}$ is closed under all four arithmetic operations *provided* that we do not divide by 0. The set $\mathbb{R} \setminus \mathbb{Q}$ is not closed under *any* arithmetic operation:

$$\sqrt{2} - \sqrt{2} = 0$$
$$(2 - \sqrt{2}) + (2 + \sqrt{2}) = 4$$
$$\sqrt{2} \cdot \sqrt{2} = 2$$
$$\sqrt{2}/\sqrt{2} = 1$$

2. Let $x_j = q + \sqrt{2}/j$.

3. These sets are intervals.

4. If b is a square root of β then $-b$ is also. Equivalently, the polynomial equations $z^2 - \beta$ must have two roots.

5.

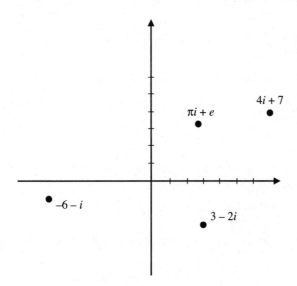

Figure 5.1s An Argand diagram.

6. First note that 1 is a cube root of this number. So $z - 1$ must divide the polynomial $p(z) = z^3 - 1$ (of which all the cube roots must be a solution). Dividing this polynomial p by $z - 1$ we find the quotient $q(z) = z^2 + z + 1$. Using the quadratic formula, we find the roots of q to be $[-1 \pm \sqrt{3}i]/2$. These are the other two cube roots of 1.

7. Divide p by $(z - \alpha)$ to obtain

$$p(z) = q(z) \cdot (z - \alpha) + r$$

Here r is the remainder, and it must be of lower degree than the divisor $(z - \alpha)$. Hence r is a constant. Now set $z = \alpha$ in this last equation. The result is

$$0 = q(0) \cdot 0 + r$$

We conclude that $r = 0$. Hence $(z - \alpha)$ evenly divides p.

8. Seeking a contradiction, we suppose that

$$\sqrt{2} + \sqrt{3} = \frac{a}{b}$$

where a, b are integers. Squaring both sides gives

$$2 + 3 + 2\sqrt{6} = \frac{a^2}{b^2}$$

Simplifying yields

$$\sqrt{6} = \frac{1}{2} \cdot \left(\frac{a^2}{b^2} - 5 \right)$$

We conclude that $\sqrt{6}$ is a rational number, and that is patently false (by the same proof that $\sqrt{2}$ is irrational).

9. Simply solve the equation

$$(x\mathbf{i} + y\mathbf{j} + z\mathbf{k} + w)^2 = 1 + \mathbf{i} + \mathbf{j}$$

Chapter 6

1. Assuming that everyone is healthy, we may suppose that the weights of people in the room range from 75 pounds to 200 pounds. That is a range of 126 possible value. And each of 300 people will have one such value. So there are 300 letters and 126 mailboxes. It follows that two people will have the same weight.

2. The waist measurements will be in the range 15 inches to 45 inches. That is a span of 31 values. But there are 50 people. Just as in the last problem, to people will have the same measurement.

3. If instead we measure waist size by millimeters, then the range will be from $15 \times 25.4 = 381$ to $45 \times 25.4 = 1143$. That is a span of 763 possible values. Since there are just 50 people, each person could have a different waist measurement in millimeters.

4. For the choice of the first 5, there are $\binom{20}{5} = 15504$ possibilities. The other three will be chosen from the remaining 15 people, and then there are $\binom{15}{3} = 455$ possibilities. The total number of ways to assign 5 and then three people to the two rooms is then $15504 \times 455 = 7054320$.

5. Given any particular denomination of card (from two through ace), there are four cards of that kind. There are four different ways to choose three from

among those. And 13 different denominations. Hence there are $4 \times 13 = 52$ different ways to form three-of-a-kind.

6. The analysis here is similar to the last problem. Given any particular denomination, there is just one way to form four of that kind. And there are 13 different denominations. Thus there are 13 different ways to from four-of-a-kind.

7. The only way to roll a 7 are $1 - 6$, $6 - 1$, $2 - 5$, $5 - 2$, $4 - 3$, and $3 - 4$ (where we are taking into account that there are two dice, so two ways to realize any particular score). And there are $6 \times 6 = 36$ possible outcomes altogether. So the likelihood is $6/36 = 1/6$.

8. The only way to get a 2 is $1 - 1$. Thus the chances of getting a 2 are $1/36$. Also there is only one way to get a 12. So the chances of getting a 12 are $1/36$. Every other value can be achieved in more than one way, so these are the only two values for which the odds are $1/36$.

9. The only ways to get a 10 are

$$1 - 3 - 6 \qquad 1 - 4 - 5 \qquad 2 - 2 - 6 \qquad 2 - 3 - 5$$
$$2 - 4 - 4 \qquad 3 - 3 - 4 \qquad 6 - 1 - 3$$

and the six permutations of each of these. So there are $6 \times 7 = 42$ rolls that give 10. There are $6 \times 6 \times 6 = 216$ possible rolls. Thus the odds are $42/216 = 7/36$.

10. $F(x) = a_0 + a_1 x + a_2 x^2 + \cdots$. Then $xF(x) = a_0 x + a_1 x^2 + a_2 x^3 + \cdots$ and $x^2 F(x) = a_0 x^2 + a_1 x^3 + a_2 x^4 + \cdots$. Then

$$F(x) - xF(x) - 2x^2 F(x) = (a_0 + a_1 x + a_2 x^2 + a_3 x^3 + \cdots)$$
$$- (a_0 x + a_1 x^2 + a_2 x^3 + \cdots)$$
$$- 2(a_0 x^2 + a_1 x^3 + a_2 x^4 + \cdots)$$
$$= a_0 + (a_1 - a_0)x + (a_2 - a_1 - 2a_0)x^2$$
$$+ (a_3 - a_2 - 2a_1)x^3 + \cdots$$
$$= a_0 + (a_1 - a_0)$$
$$= 3 - 8x$$

We conclude that

$$F(x) = \frac{3 - 8x}{1 - x - 2x^2} = \frac{3 - 8x}{-2(x + 1)(x - 1/2)}$$

We apply the method of partial fractions to this last expression to obtain

$$F(x) = \frac{11/3}{1 + x} - \frac{2/3}{1 - 2x}$$

Now expanding these expressions in geometric series as usual, we have

$$F(x) = \frac{11}{3} \sum_{j=0}^{\infty} (-x)^j - \frac{2}{3} \sum_{j=0}^{\infty} (2x)^j$$

Identifying power series coefficients, we find that

$$a_j = \frac{11}{3} \cdot (-1)^j - \frac{2}{3} \cdot 2^j$$

That is the solution of our recurrence relation.

11. Draw some pictures.

Chapter 7

1. 4×3
2. 3×5
3.

$$\begin{pmatrix} -3 & 0 & 13 \\ 9 & 5 & 15 \end{pmatrix}$$

4.

$$\begin{pmatrix} 13 & -42 \\ 22 & 9 \end{pmatrix}$$

5.

$$\begin{pmatrix} -1 & -10 & 4 \\ 1 & 2 & 9 \end{pmatrix}$$

6. We write the steps of gaussian elimination in order:

$$\left(\begin{array}{ccc|ccc} 1 & 0 & 1 & 1 & 0 & 0 \\ 0 & 1 & 1 & 0 & 1 & 0 \\ 2 & 1 & 0 & 0 & 0 & 1 \end{array}\right)$$

$$\left(\begin{array}{ccc|ccc} 1 & 0 & 1 & 1 & 0 & 0 \\ 0 & 1 & 1 & 0 & 1 & 0 \\ 0 & 1 & -2 & -2 & 0 & 1 \end{array}\right)$$

$$\left(\begin{array}{ccc|ccc} 1 & 0 & 1 & 1 & 0 & 0 \\ 0 & 1 & 1 & 0 & 1 & 0 \\ 0 & 0 & -3 & -2 & -1 & 1 \end{array}\right)$$

$$\left(\begin{array}{ccc|ccc} 1 & 0 & 1 & 1 & 0 & 0 \\ 0 & 1 & 0 & -2/3 & 2/3 & 1/3 \\ 0 & 0 & -3 & -2 & -1 & 1 \end{array}\right)$$

$$\left(\begin{array}{ccc|ccc} 1 & 0 & 1 & 1 & 0 & 0 \\ 0 & 1 & 0 & -2/3 & 2/3 & 1/3 \\ 0 & 0 & 1 & 2/3 & 1/3 & -1/3 \end{array}\right)$$

$$\left(\begin{array}{ccc|ccc} 1 & 0 & 0 & 1/3 & -1/3 & 1/3 \\ 0 & 1 & 0 & -2/3 & 2/3 & 1/3 \\ 0 & 0 & 1 & 2/3 & 1/3 & -1/3 \end{array}\right)$$

7. The rows are linearly independent so the matrix induces a mapping of $\mathbb{R}^3$ that is one-to-one and onto. So it must be invertible.

8. The probability of heads on any given flip is $0.666\ldots$ and the probability of tails is $0.333\ldots$. This is true regardless of the history of previous flips. Thus the likelihood of two heads in a row is $0.444\ldots$ and the likelihood of two tails in a row is $0.111\ldots$.

Chapter 8

1. This graph (Fig. 8.1s) does not have an Euler path because, once you go to the end of a (vertical) leg, you cannot get back.

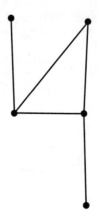

Figure 8.1s   A graph on five vertices without an Euler path.

2. This graph (Fig. 8.2s) has two distinct Euler paths because the triangle can be traversed clockwise or counterclockwise.

Figure 8.2s   A graph on five vertices with two distinct Euler paths.

3. We know that the Euler number for a sphere with one handle is 0. See Fig. 8.3s, in which $V = 1$, $E = 2$, and $F = 1$. If we add a second handle that changes the Euler number to -2. This point can be seen by examining Fig. 8.4s in which $V = 4$, $E = 10$, and $F = 4$.

4. Such a graph will have $\binom{5}{2} = 10$ edges. It will have $\binom{5}{3} = 10$ faces (because each face will be a triangle).

5. The complete graph on k vertices has $\binom{k}{2}$ edges.

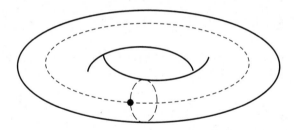

Figure 8.3s The Euler number of the sphere with one handle.

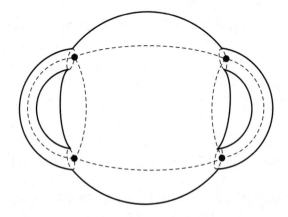

Figure 8.4s The Euler number of the sphere with two handles.

6. This graph has nine edges since each of three vertices in the first row must be connected to each of three vertices in the second row (or vice versa).

7. Of course there are five vertices. There are also five edges.

8. In Fig. 8.5s, the left-hand graph illustrates the first desideratum and the right-hand graph illustrates the second.

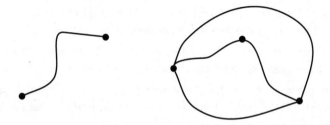

Figure 8.5s Two particular graphs.

9. The formula is $\chi = 2 - 2g$.

Chapter 9

1. (a) 2
   (b) 0
   (c) 1
   (d) 2
   (e) 2
   (f) 1

2. Let $m = m' + 2k$ and $n = n' + 2\ell$, where m' and n' are each either 0 or 1. Then

$$m \cdot n = m' \cdot n' + 2(kn' + \ell m' + 2k\ell)$$

Thus

$$(m \cdot n) \bmod 2 = m' \cdot n' = (m \bmod 2) \cdot (n \bmod 2)$$

3. Let $m = m' + 2k$ and $n = n' + 2\ell$, where m' and n' are each either 0 or 1. Then

$$m + n = m' + n' + 2(k + \ell)$$

Thus

$$(m + n) \bmod 2 = m' + n' = (m \bmod 2) + (n \bmod 2)$$

4. The prime factorization is $111 = 3 \cdot 37$ and 211 is prime. The numbers clearly have no prime factors in common.

5. We write $1024 = 2 \cdot 2 \cdot 2 \cdot 2 \cdot 2 \cdot 2 \cdot 2 \cdot 2 \cdot 2 \cdot 2$ and $100 = 2 \cdot 2 \cdot 5 \cdot 5$. Clearly the greatest common divisor is $2 \cdot 2 = 4$.

6. Plainly the sum of two 2×2 matrices is another 2×2 matrix. Matrix addition is associative just because ordinary addition of numbers is. The additive identity will be

$$\begin{pmatrix} 0 & 0 \\ 0 & 0 \end{pmatrix}$$

Given a 2×2 matrix

$$A = \begin{pmatrix} a & b \\ c & d \end{pmatrix}$$

its additive inverse will be

$$-A = \begin{pmatrix} -a & -b \\ -c & -d \end{pmatrix}$$

7. The 2×2 matrix

$$A = \begin{pmatrix} 1 & 0 \\ 0 & 0 \end{pmatrix}$$

does not have a multiplicative inverse. Hence these matrices do not form a group under multiplication.

8. We check that

$$(a \cdot b) \cdot (b^{-1} \cdot a^{-1}) = ((a \cdot b) \cdot b^{-1}) \cdot a^{-1}$$
$$= (a \cdot (b \cdot b^{-1})) \cdot a^{-1}$$
$$= (a \cdot e) \cdot a^{-1}$$
$$= a \cdot a^{-1}$$
$$= e$$

A similar calculation shows that

$$(b^{-1} \cdot a^{-1}) \cdot (a \cdot b) = e$$

It follows then that $b^{-1} \cdot a^{-1}$ is the multiplicative inverse of ab.

9. The polynomial $p(x) = x^2 + 1$ does not have a multiplicative inverse in the polynomials. So the polynomials do not form a group under multiplication.

10. (a)

$$80 = 5 \cdot 15 + 5$$
$$15 = 3 \cdot 5 + 0$$

So 5 is the greatest common divisor.

(b)

$$92 = 3 \cdot 24 + 20$$
$$24 = 1 \cdot 20 + 4$$
$$20 = 5 \cdot 4$$

Therefore 4 is the greatest common divisor.

Chapter 10

1. We use the standard transliteration $A \rightarrow 0$, $B \rightarrow 1$, and so on to transform the given message to the list of numbers

$$1\ 24\ 4\ 1\ 24\ 4\ 1\ 8\ 17\ 3\ 8\ 4$$

Now we perform the linear transformation $P \mapsto P - 3$, applied modulo 26. We obtain

$$24\ 21\ 1\ 24\ 21\ 1\ 24\ 6\ 14\ 0\ 5\ 1$$

Again the transliteration now yield the encrypted message

$$Y\ V\ B\ Y\ V\ B\ Y\ G\ O\ A\ F\ B$$

2. The coded message transliterates to the sequence of numbers

$$4\ 0\ 23\ 0\ 25\ 18\ 13\ 12\ 13\ 10$$

The decryption algorithm, applied modulo 26, yields now the sequence

$$18\ 14\ 11\ 14\ 13\ 6\ 1\ 0\ 1\ 24$$

The usual transliteration converts this to

$$\text{SOLONGBABY}$$

Remembering that we need to insert spaces and punctuation, we finally retrieve the message

$$\text{SO LONG BABY}$$

3. We notice that R occurs five times in the encrypted message. Since E is the most commonly occurring letter in the English language, we guess that E has been encoded as R. Thus we guess that the shift encryption being used here is $P \mapsto P + 13$. Thus the decryption algorithm is $P \mapsto P - 13$. We transliterate the encrypted message as usual to

$$25 \ 17 \ 17 \ 6 \ 25 \ 17 \ 20 \ 17 \ 4 \ 17$$

Now we perform the decryption to obtain

$$12 \ 4 \ 4 \ -7 \ 12 \ 4 \ 7 \ 4 \ -9 \ 4$$

Finally, this string of numbers transliterates to

MEETMEHERE

Adding spacing as usual gives the message

MEET ME HERE

4. We transliterate the message to

$$7 \ 4 \ 11 \ 11 \ 14 \ 12 \ 24 \ 7 \ 14 \ 13 \ 4 \ 24$$

Applying the affine encryption scheme (modulo 26 as usual) gives the result

$$6 \ 23 \ 18 \ 18 \ 1 \ 21 \ 5 \ 6 \ 1 \ 24 \ 23 \ 5$$

This finally becomes the encrypted message

GXSSBVFGBYXF

5. Under the usual transliteration, the message becomes

$$17 \ 3 \ 16 \ 24 \ 15 \ 7 \ 25 \ 24 \ 3 \ 16 \ 24 \ 15$$

The affine decryption scheme transforms this to

$$2 \ 0 \ 13 \ 3 \ 24 \ 8 \ 18 \ 3 \ 0 \ 13 \ 3 \ 24$$

(Notice that we have had to do some division modulo 26.) Finally, this transliterates to

CANDYISDANDY

Inserting spaces as usual gives the message

CANDY IS DANDY

6. The digraphs are

TH IS WA SN OT TH EN DX

Notice how we have added an X on the end so that the digraphs come out even. These digraphs correspond to the pairs of numbers

(19,7) (8,18) (22,0) (18,13) (14,19) (19,7) (4,13) (4,23)

Now, according to the algorith in the text, these pairs correspond to

501 226 572 481 383 501 117 127

Now we encrypt this list of numbers as

158 9 371 98 480 158 358 478

This tranlates into the roman alphabet as

GC AJ OH DU SM GC NU SK

In other words, our encrypted message is

GCAJOHDUSMGCNUSK

7. The standard transliteration of the given message is

13 14 22 8 18 19 7 4 19 8 12 4

Application of the encryption algorithm then yields

0 6 24 16 0 4 18 6 4 16 4 6

Chapter 11

1. Imitating the example in the text, this reduces to

$$[\bar{a} \times \bar{b}] + [\bar{b} \times \bar{c}] + [\bar{a} \times \bar{c}]$$

2. We write

$$a \times (a + b) = (a \times a) + (a \times b)$$
$$= a + (a \times b)$$

But if we remember that $a \times b$ is the intersection of a and b, then $a \times b$ is a subset of a. So this last line must be (remembering that $+$ is union) just a itself.

4. This is just a boolean rendition of the familiar fact

$$^c(A \cup B) = {}^cA \cap {}^cB$$

6. Similar to Sol. 2.

7. Interpret in the language of intersection and union, and then the assertion is clear.

Chapter 12

1. 0
2. 0
3. For $\epsilon > 0$, let $j > 1/\epsilon - 7$.
4. For $\epsilon > 0$, let $j > \log_{10}(1/\epsilon)$.
5. 1
6. 1/2
7. 1
8. 0
9. 0
10. 1
11. 3

Chapter 13

1. $5,\ 5 + 5/2,\ 5 + 5/2 + 5/4,\ 5 + 5/2 + 5/4 + 5/8$
2. $1/3,\ 1/3 + 1/9,\ 1/3 + 1/9 + 1/27,\ 1/3 + 1/9 + 1/27 + 1/81$
3. Diverges
4. Converges to 1
5. The terms, for $j > 10$, are smaller than $2/j^2$. So the series converges.
6. The terms are smaller in absolute value than 2^{-j}. So the series converges.
7. The sum is 8/7.
8. The sum is 7/4.
9. The sum is $\dfrac{3^7 - 1}{3^7 - 3^6}$.
10. The sum is $11^4 \cdot \dfrac{11^{10} - 1}{11 - 1}$.
11. Converges
12. Diverges
13. The ratio is $1/(j + 1)$, which tends to 0. So the series converges.
14. The root is $3j/(j + 1)$, which tends to 3. So the series diverges.

Bibliography

[ADA] J. F. Adams, On the non-existence of elements of Hopf invariant one, *Annals of Math.* 72(1960), 20–104.

[BAR] J. Barwise, ed., *Handbook of Mathematical Logic*, North-Holland, Amsterdam, 1977.

[BMS] G. Birkhoff and S. MacLane, *A Survey of Modern Algebra*, 5th ed., A.K. Peters, Wellesley, Mass., 1997.

[BLU] M. Blum, How to prove a theorem so no one else can claim it, *Proc. International Congress of Mathematicians* (Berkeley, Calif., 1986), 1444–1451, AMS, Providence, R.I., 1987.

[BSMP] M. Blum, A. De Santis, S. Micali, and G. Persiano, Noninteractive zero-knowledge, *SIAM J. Computing* 20(1991), 1084–1118.

[BG] M. Blum and S. Goldwasser, An efficient probabilistic public-key encryption scheme which hides all partial information, *Advances in Cryptology* (Santa Barbara, Calif., 1984), 289–299, *Lecture Notes in Computer Science* 196, Springer-Verlag, Berlin, 1985.

[BOM] R. Bott and J. Milnor, On the parallelizability of the spheres, *Bull. Am. Math. Soc.* 64(1958), 87–89.

[CUT] N. Cutland, ed., *Nonstandard Analysis and Its Applications*, Cambridge University Press, Cambridge, England, 1988.

[DAN] G. B. Dantzig, Programming of interdependent activities. II. Mathematical model, *Econometrica* 17(1949), 200–211.

[FFP] U. Feige, A. Fiat, and G. Persiano, Noninteractive zero-knowledge proof systems, *Advances in Cryptology—CRYPTO '87* (Santa Barbara, Calif., 1987), 52–72, *Lecture Notes in Computer Science* 293, Springer-Verlag, Berlin, 1988.

[FFS] U. Feige, A. Fiat, and A. Shamir, Zero-knowledge proofs of identity, *J. Cryptology* 1(1988), 77–94.

[JOH] P. T. Johnstone, *Stone Spaces*, Cambridge University Press, Cambridge, 1986.

[HER] I. N. Herstein, *Topics in Algebra*, Xerox College Publishing, Lexington, Mass., 1975.

[KRA1] S. G. Krantz, *The Elements of Advanced Mathematics*, 2d ed., CRC Press, Boca Raton, Fla., 2002.

[LIN] T. Lindstrøm, An invitation to nonstandard analysis, in *Nonstandard Analysis and Its Applications*, N. Cutland, ed., Cambridge University Press, Cambridge, England, 1988.

[NEL] E. Nelson, *Predicative Arithmetic*, Princeton University Press, Princeton, N.J., 1986.

[SIN] S. Singh, *Fermat's Enigma*, Anchor Books, New York, 1998.

[SUP] P. Suppes, *Axiomatic Set Theory*, Dover Publications, New York, 1972.

[WHR] A. N. Whitehead and B. Russell, *Principia Mathematica*, Cambridge University Press, Cambridge, England, 1910.

[WOD] M. K. Wood and G. B. Dantzig, Programming of interdependent activities. I. General discussion, *Econometrica* 17(1949), 193–199.

INDEX